C0-ATQ-458

HD 9750.5 .I54 1987
**International Symposium on World
Trade in Forest Products (3rd : 1987 :
University of Washington)**
Forest products trade : market trends

Forest Products Trade:
Market Trends and Technical Developments

FOREST PRODUCTS TRADE: MARKET TRENDS AND TECHNICAL DEVELOPMENTS

Edited by Jay A. Johnson and W. Ramsay Smith

University of Washington Press

SEATTLE AND LONDON

HD
9750.5
.I54
1987

Copyright © 1988 by the University of Washington Press

Printed in the United States of America

All rights reserved. No part of this publication may be reproduced or transmitted in any form or by any means, electronic or mechanical, including photocopy, recording, or any information storage or retrieval system, without permission in writing from the publisher.

The following chapters were written and prepared by U.S. government employees on official time, and are therefore in the public domain: "U.S. Hardwood Trade in the Pacific Rim" by Philip A. Araman and "The Role of Technology in Improving the Competitive Position of the U.S. Forest Products Industry" by H. M. Montrey and Jay A. Johnson.

Library of Congress Cataloging-in-Publication Data

Forest products trade : market trends and technical developments / edited by Jay A. Johnson and W. Ramsay Smith.
   p. cm. -- (Contribution / Institute of Forest Resources ; no. 61)
   Papers presented at 3rd symposium held in Mar. 1987 and sponsored by CINTRAFOR.
   Includes bibliographies and index.
   1. Forest products industry--Congresses. I. Johnson, Jay A., 1941-   . II. Smith, W. Ramsay. III. CINTRAFOR   IV. Series: Contribution (University of Washington. Institute of Forest Resources) ; no. 61.
SD12.W33 no. 61
[HD9750.5]
634.9 s--dc19
[382'.41498]
ISBN 0-295-96682-3

                                                   88-10604
                                                      CIP

Proceedings of the Third International Symposium on World Trade in Forest Products, held at the University of Washington, March 18-20, 1987, and sponsored by the University of Washington's College of Forest Resources and Center for International Trade in Forest Products (CINTRAFOR).

Contents

Preface

The mid-1980s marked a period of resurging trade in forest products on a global scale. In keeping with its mission to disseminate current and timely information about international forest products trade, CINTRAFOR sponsored the third symposium on "International Trade in Forest Products" in March 1987. This volume contains the presentations of that symposium.

Three related themes encompass the major issues and concerns evident in the reawakening of expanded forest products trade in 1987. Following the period of global economic recession and depressed markets of the first half of the 1980s, a new feeling of guarded optimism and international competition emerged. A structured consideration of trade actions and competitive reaction was needed, reflecting the reality of a truly integrated timber economy as a replacement for simple bilateral trade. Second, the relationship between technical processes and products and the success of trade was recognized, leading to the need for a much better understanding of changing technology, product standards, and product mix. Finally, the rapidly changing role of the tropical countries of Latin America and Asia as wood producers and consumers was seen as a significant factor in the overall determination of global trade patterns and competitiveness.

As was true of previous symposia, the authors of the papers contained in this volume were selected to provide a balanced overall perspective to issues pertaining to worldwide forest products trade, to focus on specific topics and regions (or countries), and to highlight significant developments in trade policies and practices. It was felt that it was important to document these developments and changes in order to provide a clear background for future trade development. Thus Latin America—because of its ability to become a major source of wood supply as well as its needs that imply a large potential market for wood products—is covered in several papers. Wood flow to North Africa and the Middle Eastern countries is also considered, as are countries that have strong manufacturing capabilities but little or no internal domestic wood supply—Korea and Taiwan. Emphasis on emerging markets is reflected in current views on the role of the People's Republic of China as a wood consumer.

Technical considerations are easily overlooked in the examination of trade flows and international trade policy. The blurring of technical and product details, however, does not diminish their importance. Consequently, contributers to this volume were asked to evaluate the impact on world trade of such matters as the role of technology, measuring and grading lumber, product standards and building codes, development of new products, and the evolving status of nonresidential construction.

Tropical countries face many unique problems—including replenishment of the resource—before they can fully utilize their extensive timber resources, and before they can effectively make use of wood to meet major social needs. Discussion of some of the problems and issues of resource and industrial development in tropical countries concludes this volume.

Papers presented at this international symposium and published here focus on solid-wood products. A companion volume presents papers related to global pulp and paper markets and trade. CINTRAFOR and the College of Forest Resources are pleased to provide these thoughtful papers on the changing dimensions of international forest product trade.

Acknowledgments

Conferences and symposia do not occur spontaneously and the proceedings do not appear out of thin air. We thank the authors for their valuable work: for their excellent presentations at the conference and for the final papers contained in this volume. These contributions form the substance of this Forest Products Trade symposium.

Many others, however, were instrumental in transforming ideas, concepts, and plans into the substance. To these people we are gratefully indebted: Dr. Gerard Schreuder and Dr. James Bethel, who served on the organizing committee; Mr. Frank Michiels of Michiels International, who offered valuable assistance, advice, and contacts; Dr. Thomas Waggener, Director of CINTRAFOR, who was instrumental in providing CINTRAFOR backing and who served an overseeing function; and Dr. David Thorud, Dean of the College of Forest Resources.

Diana Perl and her staff at the Office of Continuing Education provided flawless logistical support. Beverly Anderson and the staff of the Institute of Forest Resources organized efforts to enable the publication of the proceedings. Margaret Lahde coordinated the word-processing preparation of manuscripts. Finally we wish to thank Leila Charbonneau for her superb technical editing contribution.

Jay A. Johnson and Ramsay W. Smith

Trade Actions and Reactions

Chapter 1

An Overview of Forest Products Trade in Latin America and the Caribbean Basin

JAN G. LAARMAN, GERARD F. SCHREUDER, and ERIK T. ANDERSON

FOREST RESOURCES, PRODUCTION, AND CONSUMPTION

The forests of Latin America and the Caribbean Basin are vast in some areas and sparse in others, more or less in inverse relation to population densities. Forest types are many, unevenly distributed, and highly variable in biological complexity. Forest science and forestry practice are respectable and respected in a few countries, and weak or nonexistent in most others. Timber-based processors range in size and technology from very small sawmills to highly sophisticated pulp and paper mills. The nearly forty countries vary in size and population from tiny (especially in the Caribbean Basin) to large (e.g., Brazil and Mexico). A wide spectrum of political and socioeconomic organization reflects an equally wide spectrum of historical and cultural antecedents.

For the reasons briefly outlined above, it is usually difficult and generally misleading to summarize the production, consumption, and trade of forest products in Latin America as a single region. The analysis of production and trade will recognize seven subregions (Table 1), modifying slightly a division proposed by the Inter-American Development Bank (McGaughey and Gregersen 1983). Yet even a seven-part division masks considerable intraregional variation.

A useful synthetic description of the forest resources and industries of Latin America can be developed by combining observations from Gregersen and Contreras (1975), Erfurth and Rusche (1978), Lanly (1982), McGaughey and Gregersen (1983), Yoho and Gallegos (1984), and Zobel (1987). Here we liberally paraphrase from these sources.

Forest Resources

Forest Classification. The natural forests of Latin America can be classified in four broad types: coniferous, temperate hardwood, tropical dry hardwood (including wooded savannas), and tropical moist hardwood. Less than 3% of the natural forests are coniferous. The bulk of the forest area comprises mixed tropical hardwoods in wet or moist zones, dominated by the extensive forests of the Amazon Basin. Other areas supporting significant areas of mixed tropical wet or moist hardwoods include parts of southern Mexico, the Atlantic coast of Central America, the Guiana uplands, the Atlantic coast of Brazil, and the Pacific coast of Ecuador and Colombia (Gregersen and Contreras 1975, pp. 11-14; Lanly 1982; Yoho and Gallegos 1984, pp. 46-48).

Utilization. Utilization of the natural forests tends toward two extremes. On the one hand, vast expanses of mixed tropical hardwoods are still considered "unproductive" in the context of current technologies, infrastructure, and markets. On the other, commercially preferred species in accessible areas have been and continue to be "mined," with little chance of near-term replacement (Erfurth and Rusche 1978; McGaughey and Gregersen 1983, pp. 34-35).

Overcutting. The relatively limited areas of natural coniferous forests have been heavily overcut. In particular, the araucaria timber of southern Brazil has been largely depleted. The same ap-

Table 1. Countries of Latin America and the Caribbean Basin classified by subregion.

Subregion	Countries	
Mexico		
Central America	Belize	Honduras
	Costa Rica	Nicaragua
	El Salvador	Panama
	Guatemala	
Caribbean Islands	Bahamas	Haiti
	Barbados	Jamaica
	Bermuda	Leeward and Windward Islands
	Cayman	Martinique
	Cuba	Netherlands Antilles
	Dominica	Saint Vincent
	Dominican Republic	Trinidad and Tobago
	Guadeloupe	Turks and Caicos Islands
Northern South America	French Guiana	Suriname
	Guyana	
Brazil		
Andean Zone	Bolivia	Peru
	Colombia	Venezuela
	Ecuador	
Southern Cone	Argentina	Paraguay
	Chile	Uruguay

Note: A few small Caribbean countries are not included in this study.

plies for the natural pine stands of the Caribbean islands. Mexico and parts of northern Central America still retain pine forest, but the future of many of these forests is uncertain in the face of rapid land use changes. Chile and Argentina possess limited areas of lesser-known conifers (e.g., alerce and podocarps), but these species do not support a large industrial base (Lanly 1982; Yoho and Gallegos 1984, pp. 46-47).

Reasons for Overcutting. Uncontrolled forest cutting is blamed on many factors, varying by country, but typically including inadequate definition of forest ownership and tenure, land competition from agriculture and cattle grazing, political and economic inability to assimilate forest squatters in alternative livelihoods, poorly conceived and impractical forest laws and regulations, weak and underfunded government agencies, and lack of forestry tradition and know-how.

Forest Expansion. In a few subregions (principally Brazil and the Southern Cone) the area of forest planting has been expanding rapidly through the last two decades. Although most of these plantations are on private lands, they have been generously subsidized by government tax, cost-sharing, or credit programs. Various observers regard the plantations as "compensatory plantings" which will increasingly relieve cutting pressure on the natural forests (Zobel 1987).

Plantations. Already plantations provide about one-third of Latin America's total industrial roundwood production, even though they occupy less that 1% of the productive forest area. The Inter-American Development Bank projects that by the year 2000, timber cut from plantations will be almost four times the current level, accounting for half of total roundwood production. Plantation harvests will be heavily concentrated in Brazil and the Southern Cone (McGaughey and Gregersen 1983, pp. 37-38).

World Markets. In various parts of South America and especially Brazil, the delivered costs of plantation timber are sufficiently low to compete favorably in world markets. In the leading forest products companies, management and technology have become fairly sophisticated. Research is being conducted and then applied. Several large Brazilian pulp mills are modern and ideally sited. Extensive areas of flat and accessible savannas, pastures, and brushlands remain available for tree planting at relatively low cost. Rotations are short and getting shorter. Silviculture and genetic engineering are producing dramatic increases in the productivity of hardwood plantations (e.g., eucalypts and gmelina). Regarding tropical and subtropical pines, the high proportion of juvenile wood is expected to become less of a utilization problem with the increasing use of technologies like thermomechanical pulping, chemithermomechanical pulping, and oriented strandboard (Zobel 1987).

Production and Consumption

Wood Production. With approximately 8% of the world's population, Latin America and the Caribbean Basin account for disproportionately smaller shares of forest-based production (Table 2). Yet while these shares are small in absolute terms, they have been growing rapidly. Between 1974 and 1984 the region's shares of world production expanded as follows: industrial roundwood, 64%; sawnwood and sleepers, 46%; wood-based panels, 40%; woodpulp, 86%; and paper and paperboard, 31%.

Exports and Imports. At a highly aggregate level, Table 2 also shows regional shares and trends for exports and imports. As with production, the region's share of world trade is lower than its share of world population. Relative to the rest of the world, there is little trading of unprocessed logs (i.e., industrial roundwood). From very small bases in 1974, export shares have expanded dramatically in all major product groups except sawnwood and sleepers. In comparison, import shares were static or falling in most major product groups. When looking for explanations, it is im-

Table 2. The share of Latin America and the Caribbean Basin in world production, exports, and imports of forest products (1974, 1980, 1984).

Product Group	% of World Total (quantity basis)		
	1974	1980	1984
Industrial roundwood			
Production	3.9	6.8	6.4
Exports	0.4	1.0	1.0
Imports	0.2	0.2	0.2
Sawnwood and sleepers			
Production	3.9	5.7	5.7
Exports	3.2	3.6	2.5
Imports	3.2	4.0	2.7
Wood-based panels			
Production	3.0	4.2	4.2
Exports	2.0	3.8	3.6
Imports	1.3	3.0	2.4
Woodpulp			
Production	2.2	3.8	4.1
Exports	1.6	6.2	6.9
Imports	4.7	3.8	3.3
Paper and paperboard			
Production	3.5	4.5	4.6
Exports	0.8	1.2	2.5
Imports	7.4	7.1	5.1

Source: FAO *Yearbook of Forest Products* 1973-84.

portant to take into account the recent macroeconomic crises affecting most countries of the region (see the following section on "Recent Performance of the Forest-Based Industries").

Products. Compared with North America, a relatively large share of Latin America's consumption of solid-wood products is in the form of reconstituted panels, flooring, and manufactured products like furniture. Regarding pulp and paper products, a relatively large share of the region's consumption is in newsprint, with lesser shares in industrial packaging and paperboard (Yoho and Gallegos 1984, pp. 49-50).

Consumption of Wood Products. Among other factors, low per capita consumption of wood products in Latin America and the Caribbean Basin is customarily ascribed to preferences for masonry over wood in construction, lagging incomes for a significant proportion of the population, and high import barriers erected to protect fledgling forest-based industries (Table 3).

Major Consuming Countries. Consumption of forest products is highly skewed in the direction of three geographically large countries with large national incomes. Brazil, Mexico, and Argentina together account for half or more of the total region's consumption in every major forest product group. Table 4 shows the consumption shares of these three countries for 1984.

Importing Countries. Most individual countries have been and continue to be net importers of wood products on a value basis (Table 5). The two principal exceptions are Brazil and Chile. Paraguay, Honduras, Guyana, and French Guiana also are net exporters, but on a much reduced scale. In 1984 the greatest sectoral trade deficits were recorded by Mexico, Venezuela, Cuba, Argentina, Trinidad and Tobago, and Ecuador—each experiencing deficits exceeding U.S.$100 million.

Table 3. Nonweighted average tariff rates on forest products in selected countries of Latin America and the Caribbean Basin (1970s).

Country	Roundwood (%)	Primary Products (%)	Secondary Products (%)
Mexico (1979)	13.0	14.2	34.7
Caribbean Islands			
Bahamas (1977)	24.1	29.8	31.5
Jamaica (1976)	13.7	25.4	33.6
Northern South America			
Suriname (1971)	23.7	27.7	NA
Brazil (1977)	43.0	56.4	65.3
Andean Zone			
Bolivia (1977)	11.1	33.3	36.2
Colombia (1977)	13.9	40.2	52.1
Ecuador (1977)	20.0	46.4	58.0
Peru (1977)	NA	NA	60.0
Venezuela (1977)	20.0	81.0	81.2
Southern Cone			
Argentina (1979)	15.5	18.4	29.1
Paraguay (1978)	28.9	25.7	25.3

NA = Not available.
Source: UNIDO (1983).

Table 4. Apparent consumption (production + imports − exports) of forest products in Brazil, Mexico, and Argentina in relation to totals for all of Latin America and the Caribbean Basin (1984).

Product Group	Brazil	Mexico	Argentina
Industrial roundwood			
1,000 m³	57,675	7,287	4,977
% of regional total	62.6	7.9	6.2
Sawnwood and sleepers			
1,000 m³	15,656	2,000	1,558
% of regional total	60.3	7.7	6.0
Wood-based panels			
1,000 m³	2,132	760	378
% of regional total	49.8	17.8	8.8
Woodpulp			
1,000 m.t.	2,478	703	630
% of regional total	32.2	9.1	8.2
Paper and paperboard			
1,000 m.t.	3,281	2,476	1,031
% of regional total	33.7	25.5	10.6

Source: FAO *Yearbook of Forest Products* 1973-84.

RECENT PERFORMANCE OF THE FOREST-BASED INDUSTRIES

Macroeconomic Context

At the beginning of the 1980s the cyclical but continuously positive economic growth which most of Latin America had experienced over the previous thirty years came to an abrupt and distressing end. The region's economic depression has been the longest, most widespread, and most severe since the Great Depression. Practically all countries and sectors have been affected, including the forest-based industries. A brief consideration of this macroeconomic distress prefaces an examination of recent trade flows.

Heavy spending and consumption throughout most of the 1970s led to balance-of-payments deficits and rapidly rising prices. Because of unprecedented permissiveness in international financial circles, the inflow of external credit to sustain high consumption was generous and easy (Prebisch 1985). By the end of the 1970s the recession in the industrialized countries had grown more serious. International interest rates rose quickly, prices of Latin America's traditional exports fell, and the risks of lending and borrowing in a recessive context became obvious. By 1982-83 the inflow of new capital to the region had contracted dramatically (IDB 1985, pp. 17-31; Gonzalez 1985).

The highly stressed period 1981-83 was characterized by a sharp reduction in the formation of gross domestic product (GDP), a dramatic reduction in rates of investment, a marked increase in capital flight, a pronounced intensification of monetary inflation, a disturbing rise in unemployment and underemployment, and a painful reduction in real wages. In addition, the sectors dependent on international trade were hurt by a deterioration in terms of trade, a loss of foreign exchange reserves, and a marked increase in external debt servicing commitments (Iglesias 1984).

Beginning in 1984, there was a pause and in some cases a reversal of many downward trends (IMF 1985, p. 1). Yet the recovery is still inadequate. Economic projections show that, even under optimistic assumptions, several years will be required to recover welfare levels already attained before the depression began. The United Nations Commission for Latin America and the Caribbean (ECLAC) estimates that growth rates of 5-6% annually are needed if average income per capita in 1990 is to reach the level that prevailed in 1980.

Table 5. Trade balances in all forest products (exports – imports), Latin America and the Caribbean Basin, by subregion and country (1974, 1980, 1984).

Subregion	Trade Balances (million U.S.$)		
	1974	1980	1984
Mexico	–181.2	–598.4	–364.6
Central America	–74.5	–180.0	–156.5
Belize	1.5	–1.5	–1.0
Costa Rica	–45.0	–38.6	–49.6
El Salvador	–12.5	–28.9	–24.5
Guatemala	–5.9	–71.7	–45.9
Honduras	14.7	3.0	10.2
Nicaragua	–4.2	–9.1	–9.8
Panama	–23.1	–33.2	–35.9
Caribbean Islands	–188.0	–469.2	–502.0
Bahamas	–3.1	–7.5	–7.8
Barbados	–7.0	–24.3	–19.7
Cuba	–77.0	–200.9	–188.9
Dominica	--	--	–0.9
Dominican Republic	–23.0	–71.4	–58.8
Guadeloupe	–3.0	–6.3	–14.2
Haiti	–2.3	–4.8	–5.0
Jamaica	–28.6	–34.1	–47.9
Martinique	–5.2	–11.3	–14.1
Netherlands Antilles	–9.9	–9.2	–9.6
Saint Vincent	--	–0.5	–3.0
Trinidad and Tobago	–28.9	–98.9	–132.1
Northern South America	–0.8	3.0	–0.1
French Guiana	0.5	2.3	1.1
Guyana	–2.4	–0.2	0.3
Suriname	1.1	0.9	–1.5
Brazil	–113.1	590.4	853.9
Andean Zone	–201.0	–401.8	–498.4
Bolivia	7.5	8.6	–6.5
Colombia	–57.7	–92.4	–64.8
Ecuador	–30.0	–66.6	–101.1
Peru	–35.4	–31.0	–40.3
Venezuela	–85.4	–220.4	–285.7
Southern Cone	–115.7	50.6	226.2
Argentina	–244.4	–395.2	–168.5
Chile	124.5	413.7	321.9
Paraguay	18.5	54.9	79.8
Uruguay	–14.3	–22.8	–7.0
All countries	–874.5	–1,005.4	–441.5

Source: FAO *Yearbook of Forest Products* 1973-84.

Moreover, the 1984 improvements recorded in some countries have been set back or erased by recent price shifts. The 1985-86 drop in petroleum prices illustrates the extreme fragility of the recovery process for countries like Mexico, Venezuela, Ecuador, and other petroleum exporters.

Impact on Production Levels

The widespread economic turmoil in Latin America has had a severe impact on the forest-based industries of some countries. Yet elsewhere within the region the forest-based industries appear to have been expanding. Table 6 reports the major changes in forest-based production for the period 1980-83. The year 1980 preceded the worst difficulties, while 1983 marked the trough of the depression for most of the region. Therefore, the table refers to a four-year period in which there were pronounced downward pressures across most economies. Countries not individually listed in the table showed neither a clear increase nor a clear decrease for 1980-83.

Table 6. Major production shifts in the forest-based industries of Latin America and the Caribbean Basin between 1980 and 1983.

Increase/Decrease by Country	% Change (quantity basis) by Product Group			
	Sawnwood and sleepers	Wood panels	Wood-pulp	Paper and paperboard
Major Increases				
Mexico		+26		
Jamaica				+80
Costa Rica				+50
Panama				+115
Brazil			+16	
Ecuador		+34		+31
Argentina	+32		+54	+21
Major Decreases				
Mexico	−23			
Belize	−20			
Costa Rica	−28		−40	
Guatemala		−33		−50
Honduras	−14			
Nicaragua	−45			
Suriname	−23			
Bolivia	−56			
Colombia	−26			
Peru		−32		
Venezuela	−40			
Argentina		−18		
Chile	−26			
Uruguay	−84	−25		−17

Note: "Major" increase or decrease equals 10% or more.
Source: FAO *Yearbook of Forest Products* 1973-84.

Table 6 indicates that the solid-wood industries were more vulnerable than were the pulp and paper industries to the economic downturn. This is not unexpected in view of the usual demand patterns in cyclical versus steady-growth markets. At one extreme, sawnwood production increased in one country but contracted in eleven. In contrast, paper production increased in five countries but contracted in only two. Pulp production expanded in two large countries and contracted in only one small one.

Several countries posted production gains in some industries and losses in others. Shifts in other countries were more uniform. For example, the forest-based industries of Uruguay realized unambiguous net contraction during the period in question. The opposite appears to have occurred in Ecuador.

For Latin America and the Caribbean Basin as a whole, the four largest producers of forest products are Brazil, Chile, Mexico, and Argentina. Brazil's pulp production expanded modestly during the period in question. Chile's sawnwood industry contracted. Mexican sawnwood production fell, while Mexican panel production rose. Argentina recorded production gains for sawnwood, pulp, and paper; production declines occurred for wood panels.

Impact on Exports and Imports

Tables 7 and 8 summarize changes in the total value of forest product exports and imports. Although countries like Brazil showed increases in export volumes for each of panels, pulp, and paper, Brazil's net change in export value for 1980-83 was slightly negative. To the extent that the figures are reliable, they imply a softening of export prices. FAO's tables on unit values indicate that Brazil's f.o.b. unit value for pulp exports fell by approximately 23% between 1980 and 1983.

Table 7. Shifts in total value of forest product exports in Latin America and the Caribbean Basin between 1980 and 1983 (% change).

Exporting Country	Increased	Relatively Static	Decreased
Mexico	+68		
Central America			
Belize		−19	
Costa Rica		−8	
El Salvador	+615		
Guatemala			−37
Honduras		−1	
Nicaragua		0	
Panama		−2	
Caribbean Islands			
Jamaica	+108		
Trinidad and Tobago			−51
Northern South America			
French Guiana		+11	
Guyana		0	
Suriname		−2	
Brazil		−4	
Andean Zone			
Bolivia			−71
Colombia	+74		
Ecuador	+22		
Peru		+4	
Southern Cone			
Argentina			−32
Chile			−32
Paraguay		−2	
Uruguay			−35

Source: FAO *Yearbook of Forest Products* 1973-84.

Table 8. Shifts in total value of forest product imports in Latin America and the Caribbean Basin between 1980 and 1983 (% change).

Importing Country	Increased	Relatively Static	Decreased
Mexico			−40
Central America			
Belize			−23
Costa Rica	+ 28		
El Salvador		+4	
Guatemala			−33
Honduras		+9	
Nicaragua		0	
Panama		−16	
Caribbean Islands			
Bahamas		−5	
Barbados		0	
Cuba		−13	
Dominican Republic		+12	
Guadeloupe	+124		
Haiti		+3	
Jamaica	+160		
Martinique	+ 24		
Netherlands Antilles		−3	
Trinidad and Tobago	+ 33		
Northern South America			
French Guiana			−44
Guyana		−7	
Suriname	+ 20		
Brazil			−39
Andean Zone			
Bolivia		+5	
Colombia		−3	
Ecuador	+ 44		
Peru	+ 78		
Venezuela		+17	
Southern Cone			
Argentina			−48
Chile		−4	
Paraguay			−32
Uruguay			−59

Source: FAO *Yearbook of Forest Products* 1973-84.

During the same period, the drop in f.o.b. unit value for pulp exports from Chile was 37%. Export unit values for both coniferous and nonconiferous sawnwood were rising, but neither country increased sawnwood export volumes. Consequently, export earnings (when denominated in dollars) by the region's two largest exporters were flat.

Several smaller exporters recorded large increases in export value, while others showed large decreases. The export earnings of a third and somewhat larger group of small exporters remained relatively static. Many of these changes undoubtedly are explained by short-term aberrations and statistical anomalies (Durst et al. 1986). Even if real, these changes do not have much significance for regional totals or averages. Yet they may be critical at national and local levels.

Latin America's five largest importers of forest products are Mexico, Venezuela, Cuba, Argentina, and Brazil (see following section). Of these five, only Venezuela registered a modest increase in forest product imports during the four-year period (Table 8). Reductions in forest product imports of almost one-half occurred in Argentina, Mexico, and Brazil. These three large countries were among those most seriously constrained by pressures to correct balance-of-payments problems.

Among the smaller importers, both large increases and large decreases were observed in roughly equal numbers. A third and slightly larger number of small importers recorded only slight or no change. This distribution is similar to the pattern previously noted for the small exporters. The import changes are calculated from a small base, with only minor consequences for regional trade balances. Yet these changes may have been important within individual small countries.

LEADING EXPORTERS AND IMPORTERS

Table 9 lists the ten leading exporters and importers of all forest products, with ranking based on figures for 1984. The year 1984 was the latest for which complete FAO data are available. Tables 10 to 14 supplement Table 9 by itemizing the main exporters and importers in each major product group. Average annual growth rates to 1984 are computed for two subperiods, 1974-80 versus 1980-84. The first subperiod corresponds to years of buoyant trade expansion, while the second overlaps the period of recent economic depression.

Referring to Table 9, the pace of export growth in Latin America and the Caribbean Basin exceeded the world average during both subperiods, especially during the first one. The region's import growth paralleled that of the world as a whole until 1980, after which time ensuing import contractions were relatively more severe in Latin America (–5% in the region compared with a world average of –2%).

Top Exporters

Brazil and Chile consistently were the top exporters in 1974, 1980, and 1984 (Table 9). Their combined share of regional export earnings was 68% in 1974, increasing to 82% in 1980 and 1984. These two countries were among the top five exporters in every product group, and the only two exporters of woodpulp (Tables 10-14).

Since 1980 Brazil has shown remarkable earnings growth in exports of paper and paperboard, modest earnings growth in exports of wood-based panels and woodpulp, and negative earnings growth in exports of sawnwood. Between 1980 and 1984, Chile's export earnings from logs and sawnwood dropped substantially; its export earnings in other categories did not increase sufficiently to offset this decline.

However, a few qualifications are in order. The figures in Tables 10 to 14 are expressed in current dollars. Hence they overstate the growth of export purchasing power, particularly since the early 1980s were years of relatively high price inflation in the United States. Second, the tables cannot account for the multiple dollar exchange rates (official, parallel, financial, etc.) in different countries at different times and for different buyers and sellers of dollars. Consequently, the figures in Tables 9 to 14 are not necessarily valid indicators of export earnings when considering conversions to local currencies.

FAO's statistics on direction of trade show that Brazil's markets for exports of sawnwood (mainly tropical hardwoods) are the United States, the United Kingdom, and the rest of Europe. The largest export markets for Brazilian veneer and plywood are the United States and Europe, respectively. Most exports of fiberboard flow to the United States. Brazilian woodpulp is exported to a variety of destinations, but principally to the Far East (Japan and China), Europe, and the United States. Brazil's exports of paper and paperboard (chiefly newsprint) likewise reach a large number of different markets (i.e., United States, Europe, Africa, Far East, and Latin America).

Table 9. Top ten exporters and importers of all forest products, Latin America and the Caribbean Basin, by country (1974, 1980, 1984).

Country	1974	1980	1984	Average Annual Growth	
				1974-80	1980-84
	(million U.S.$)			(simple %)	
Top Exporters					
1. Brazil	236	864	1,029	44	5
2. Chile	131	459	371	42	-5
3. Paraguay	23	66	87	31	8
4. Colombia	7	27	49	48	20
5. Honduras	44	31	31	-5	0
6. Costa Rica	3	21	23	100	2
7. Ecuador	7	27	19	48	-7
8. Mexico	3	11	18	44	16
9. Suriname	7	12	12	12	0
10. Guatemala	12	22	10	14	-14
Total	473	1,540	1,649	37	2
All Latin America and Caribbean	537	1,615	1,699	33	1
World	29,168	55,883	50,682	15	-2
Top Importers					
1. Mexico	184	609	383	38	-9
2. Venezuela	85	220	286	26	8
3. Cuba	77	201	189	27	-1
4. Argentina	273	412	177	8	-14
5. Brazil	349	274	175	-4	-9
6. Trinidad and Tobago	29	99	132	40	8
7. Ecuador	37	94	120	26	7
8. Colombia	65	119	115	14	1
9. Costa Rica	48	60	72	4	5
10. Dominican Republic	23	71	59	35	-4
Total	1,170	2,159	1,708	14	-5
All Latin America and Caribbean	1,411	2,620	2,142	14	-5
World	33,054	62,252	55,797	15	-2

Note: Trade values were rounded to nearest million dollars after growth rates were calculated and columns added.
Source: FAO *Yearbook of Forest Products* 1973-84.

Chile's main market for exports of logs (pine) has been the Far East. Pine sawnwood is exported to the Middle East and a diversity of countries within the Latin American region. Chile's woodpulp is sold to the Far East and Europe, with the United States also taking a small share. Chile's exports of paper and paperboard (mainly newsprint) are sold to Brazil, Venezuela, and other countries within the region.

Among the medium-scale exporters, Paraguay and Colombia showed respectable earnings growth in both subperiods (Table 9). Most of Paraguay's exports are sawnwood and panels (Tables

Table 10. Top five exporters and importers of industrial roundwood, Latin America and the Caribbean Basin, by country (1974, 1980, 1984).

Country	1974	1980	1984	Average Annual Growth	
				1974-80	1980-84
		(million U.S.$)		(simple %)	
Top Exporters					
1. Chile	--	57	30	--	-12
2. Guyana	1	3	3	23	0
3. Brazil	7	1	2	-13	9
4. Suriname	1	2	1	17	-9
5. Honduras	1	4	1	48	-21
Total	10	67	37	95	-11
All Latin America and Caribbean	15	73	38	64	-12
World	4,469	8,690	5,901	16	-12
Top Importers					
1. Venezuela	--	2	16	--	141
2. Brazil	5	7	8	7	4
3. Dominican Republic	--	10	7	--	-8
4. Mexico	1	16	4	270	-19
5. Trinidad and Tobago	--	--	2	--	--
Total	6	35	36	80	<1
All Latin America and Caribbean	16	48	41	33	4
World	6,345	12,315	7,775	16	4

Note: Trade values were rounded to nearest million dollars after growth rates were calculated and columns added.
Source: FAO *Yearbook of Forest Products* 1973-84.

11 and 12) shipped to Brazil, since much of the Paraguayan forest products industry has Brazilian ownership. Colombia's exports are almost entirely paper and paperboard (Table 14).

Top Importers

The region's top five importers of forest products in 1984 also were the top five importers in 1980 and 1974, but the internal ordering shifted considerably (Table 9). Brazil and Argentina dropped from first and second positions in 1974 to fifth and fourth positions, respectively, by 1984. Brazil's expenditures for imports already were falling in the first subperiod, apparently reflecting ongoing import substitution of paper and paperboard (Table 14).

Meanwhile, Mexico's expenditures for exports more than tripled from 1974 to 1980. A large part of this increase was explained by rapid growth in the import demand for paper and paperboard (Table 14). Even with the sharp reduction in Mexico's imports since 1980, Mexico remained the leading importer of forest products in 1984.

Of the five largest importers in 1984, only Venezuela registered an increase in expenditures since 1980 (also see Table 8). The three major consumers of forest products—Brazil, Mexico, and Argentina—reduced imports dramatically. Because the import expenditures are expressed in current dollars rather than real dollars, Tables 9 to 14 understate the loss of purchasing power for exporters to these countries.

Considering the dynamics of individual product groups for the subperiod 1980-84, imports of sawnwood and sleepers fell sharply (except in Trinidad and Tobago). Regional imports of wood-

The King's Library

Table 11. Top five exporters and importers of sawnwood and sleepers, Latin America and the Caribbean Basin, by country (1974, 1980, 1984).

Country	1974	1980	1984	Average Annual Growth	
				1974-80	1980-84
		(million U.S.$)		(simple %)	
Top Exporters					
1. Brazil	127	211	145	11	−8
2. Chile	12	149	74	186	−13
3. Paraguay	21	52	62	24	5
4. Honduras	39	20	30	−8	12
5. Ecuador	7	4	16	−6	72
Total	206	437	327	18	−6
All Latin America and Caribbean	240	475	354	16	−6
World	6,258	12,339	10,646	16	−3
Top Importers					
1. Trinidad and Tobago	5	41	74	108	20
2. Cuba	51	88	70	12	−5
3. Mexico	13	67	62	68	−2
4. Argentina	84	139	51	11	−16
5. Venezuela	1	31	28	395	−2
Total	155	367	286	23	−6
All Latin America and Caribbean	205	464	360	21	−5
World	7,076	13,966	11,487	16	−4

Note: Trade values were rounded to nearest million dollars after growth rates were calculated and columns added.
Source: FAO *Yearbook of Forest Products* 1973-84.

based panels decreased slightly, but with modest increases posted by Venezuela, Cuba, and Trinidad and Tobago. Woodpulp imports declined for all large importers except Venezuela. Finally, among the top importers of paper and paperboard, only Ecuador increased import expenditures during the subperiod in question.

Regarding direction of trade, imports of coniferous sawnwood are mainly from the United States (to Mexico and the Caribbean Islands), the U.S.S.R. (to Cuba), Canada, and Chile. Plywood is imported from the Republic of Korea, Brazil, the United States, and the Soviet Union. Mexico imports woodpulp mainly from the United States and Canada; Venezuela imports woodpulp from the United States, Canada, Brazil, and Chile. Venezuela imports newsprint chiefly from Canada, while Brazil imports newsprint from Canada and also Chile. Over half of the region's paper and paperboard imports (quantity basis) are from the United States, with smaller suppliers being Brazil, the U.S.S.R., Finland, Canada, and Sweden.

TRADE WITH THE UNITED STATES

The United States plays a key role in forest products trade with the region. Table 15 shows U.S. exports and imports for 1985, the latest year for which complete data are available. Data in Table

Table 12. Top five exporters and importers of wood-based panels, Latin America and the Caribbean Basin, by country (1974, 1980, 1984).

Country	1974	1980	1984	Average Annual Growth	
				1974-80	1980-84
		(million U.S.$)		(simple %)	
Top Exporters					
1. Brazil	49	125	140	26	3
2. Paraguay	2	14	25	97	19
3. Mexico	--	7	9	--	6
4. Chile	--	7	9	--	6
5. Costa Rica	2	8	6	47	-5
Total	53	161	189	34	4
All Latin America and Caribbean	66	204	207	35	<1
World	2,422	5,157	4,639	19	-2
Top Importers					
1. Venezuela	2	22	32	173	10
2. Cuba	4	14	24	50	17
3. Trinidad and Tobago	2	15	21	140	11
4. Mexico	6	21	9	45	-15
5. Brazil	1	14	7	193	-13
Total	14	86	92	88	2
All Latin America and Caribbean	31	133	126	54	-1
World	2,595	5,234	4,706	17	-2

Note: Trade values were rounded to nearest million dollars after growth rates were calculated and columns added.
Source: FAO *Yearbook of Forest Products* 1973-84.

15 may not agree with data in Tables 9 to 14 because of the difference in years (1984 versus 1985) and sources (FAO versus U.S. Department of Commerce).

U.S. Exports

In 1985, U.S. exports to the region ($784 million) were almost twice the level of imports from the region ($400 million). The major export market was Mexico, which purchased 41% of total U.S. exports. Mexico bought 31% of U.S. exports of paper and paperboard, 40% of U.S. exports of solid-wood products, and 73% of U.S. exports of woodpulp.

After Mexico, the other principal markets for U.S. forest product exports are the Andean Zone, the Caribbean Islands, and Central America. In 1985 these three subregions accounted for 20, 18, and 15% of total U.S. exports to the region, respectively. Three of the Andean countries (Venezuela, Ecuador, and Colombia) were significant purchasers of U.S. paper and paperboard, as were five countries of Central America (Panama, Costa Rica, Honduras, Guatemala, and El Salvador). Several of the Caribbean Islands have been major purchasers of U.S. solid-wood products (Wisdom et al. 1986); in 1985 this subregion accounted for nearly half of U.S. exports in the solid-wood group. Besides Mexico, Venezuela was the only other significant purchaser of U.S. woodpulp.

Table 13. Top two exporters and top five importers of woodpulp, Latin America and the Caribbean Basin, by country (1974, 1980, 1984).

Country	1974	1980	1984	Average Annual Growth 1974-80	1980-84
		(million U.S.$)		(simple %)	
Top Exporters					
1. Brazil	37	364	396	148	2
2. Chile	78	197	200	25	<1
Total	115	561	596	65	2
All Latin America and Caribbean	115	561	596	65	2
World	5,459	9,543	8,974	12	−1
Top Importers					
1. Mexico	85	107	81	4	−6
2. Venezuela	21	59	77	31	8
3. Argentina	71	68	31	−1	−14
4. Colombia	26	30	25	2	−4
5. Cuba	6	23	18	12	−6
Total	209	287	232	6	−5
All Latin America and Caribbean	326	373	279	2	−6
World	5,801	9,781	9,318	11	−1

Note: Trade values were rounded to nearest million dollars after growth rates were calculated and columns added.
Source: FAO *Yearbook of Forest Products* 1973-84.

Only 4% of U.S. forest product exports went to Brazil, and only 1% went to the Southern Cone. Northern South America was an even smaller market.

U.S. Imports

In 1985 two countries, Brazil and Mexico, were the sources of more than 80% of U.S. forest product imports from Latin America and the Caribbean Basin. Imports from Brazil were chiefly solid-wood products ($113 million). But the United States also purchased substantial quantities of Brazilian pulp ($74 million), and limited quantities of Brazilian paper and paperboard ($19 million). Imports from Mexico were mainly solid-wood products ($96 million), with lesser imports of paper and paperboard ($24 million).

Brazil's exports of solid wood to the United States included a wide range of products, most important of which were hardwood lumber, hardboard, hardwood plywood, and hardwood veneer. A high proportion of Mexico's solid-wood exports to the United States were moldings. U.S. purchases of Brazilian pulp were mainly bleached hardwood (60%) as opposed to bleached softwood and unspecified (40%).

Other U.S. imports exceeding $5 million in 1985 included solid wood from Honduras, solid wood from Ecuador, paper and paperboard from Venezuela, solid wood from Chile, and paper and paperboard from Chile. However, none of these flows are particularly significant for the U.S. total of $400 million.

Table 14. Top five exporters and importers of paper and paperboard, Latin America and the Caribbean Basin, by country (1974, 1980, 1984).

Country	1974	1980	1984	Average Annual Growth	
				1974-80	1980-84
		(million U.S.$)		(simple %)	
Top Exporters					
1. Brazil	17	161	346	140	29
2. Chile	41	49	58	3	5
3. Colombia	1	24	49	327	27
4. Costa Rica	1	13	15	272	4
5. Guatemala	7	18	9	24	−13
Total	67	264	477	49	20
All Latin America and Caribbean	102	299	503	32	17
World	10,485	20,058	20,452	15	<1
Top Importers					
1. Mexico	79	399	227	68	−11
2. Brazil	250	191	138	− 4	−7
3. Ecuador	37	89	109	24	6
4. Argentina	115	183	88	10	−13
5. Colombia	38	82	82	19	0
Total	518	944	646	14	−8
All Latin America and Caribbean	831	1,602	1,335	15	−4
World	11,173	20,835	22,413	14	2

Note: Trade values were rounded to nearest million dollars after growth rates were calculated and columns added.
Source: FAO *Yearbook of Forest Products* 1973-84.

LOOKING TO THE FUTURE

For several years, the near-term outlook for forest product trade in Latin America and the Caribbean Basin has been highly uncertain (Yoho and Gallegos 1984). Given the seriousness of macroeconomic distress in so many countries, forest-based production and trade are governed by forces above and outside the sector and region. Several countries will continue to wrestle with high inflation, shrinking government expenditures, low capital inflow, and painful debt servicing (Laarman 1986).

Thus it is both curious and unfortunate that key outlook projections for Latin America's forest products output and trade make almost no reference to the region's recent economic difficulties. Projections to the year 2000 by the Inter-American Development Bank (McGaughey and Gregersen 1983) were made during the most troubled years for the region, yet these projections largely ignore that context. Similarly, more recent projections to the same year by FAO (1986) present a global view with little qualitative attention to macroeconomic shifts in different regions.

Although the main problems of the Latin American economies are not resolved, the macroeconomic outlook in the majority of countries is brighter than it was through 1983. Beginning in 1984, several countries have been able to negotiate more favorable conditions for debt rescheduling. Beginning in 1985-86, three exogenous developments having major consequences for

Table 15. U.S. exports (X) and imports (M) in forest product trade with Latin America and the Caribbean Basin (1985).

Subregion	Solid-wood Products		Woodpulp		Paper and Paperboard	
	X	M	X	M	X	M
	(million U.S.$)					
Mexico	76.0	96.4	108.6	--	138.1	23.6
Central America	6.6	17.3	3.8	--	106.3	0.5
Belize	0.4	0.1	--	--	0.2	0.2
Costa Rica	0.5	4.1	2.8	--	25.6	0.3
El Salvador	0.9	0.1	0.4	--	11.8	--
Guatemala	0.8	2.6	--	--	17.0	--
Honduras	0.2	8.8	--	--	24.3	--
Nicaragua	0.1	--	--	--	0.2	--
Panama	3.7	1.6	0.6	--	27.2	--
Caribbean Islands	87.8	1.8	0.4	--	55.9	0.2
Bahamas	18.4	--	--	--	1.3	--
Barbados	3.8	--	--	--	2.0	--
Bermuda	3.8	--	--	--	1.1	--
Cayman	4.0	--	--	--	0.1	--
Dominican Republic	13.2	0.1	--	--	17.7	0.2
Guadeloupe/Martinique	0.5	--	--	--	2.8	--
Haiti	4.4	1.5	--	--	3.3	--
Jamaica	10.7	0.1	0.1	--	11.5	--
Leeward/Windward Islands	8.5	--	--	--	5.4	--
Netherlands Antilles	8.0	--	--	--	1.8	--
Trinidad and Tobago	11.9	0.1	0.3	--	8.9	--
Turks and Caicos Islands	0.6	--	--	--	--	--
Northern South America	0.3	1.7	--	--	2.4	--
French Guiana	0.1	0.2	--	--	--	--
Guyana	0.1	1.1	--	--	0.4	--
Suriname	0.1	0.4	--	--	2.0	--
Brazil	7.4	112.6	0.6	74.0	24.3	19.0
Andean Zone	8.8	14.6	34.0	--	112.3	5.5
Bolivia	0.1	--	--	--	--	--
Colombia	1.8	2.4	3.8	--	23.7	0.2
Ecuador	0.1	9.5	0.5	--	34.1	--
Peru	1.7	2.6	1.2	--	2.5	0.2
Venezuela	5.1	0.1	28.5	--	52.0	5.1
Southern Cone	2.1	15.3	1.2	4.5	7.4	13.1
Argentina	1.2	2.8	1.0	--	3.5	4.8
Chile	0.8	11.6	0.2	4.5	3.5	8.3
Paraguay	--	0.9	--	--	0.2	--
Uruguay	0.1	--	--	--	0.2	--
All countries	189.0	259.7	148.6	78.5	446.7	61.9

Note: Solid-wood products include all those defined in the NFPA series cited below. The paper and paperboard category excludes paper bags, boxes, stationery, and other manufactured products.

Sources: U.S. Bureau of the Census, Foreign Trade Statistics, Schedule A (imports) and ScheduleE (exports); National Forest Products Association, Wood Products International Trade Statistics, vol. 2, 1985.

economic prospects are lower international interest rates, lower value of the U.S. dollar against many major currencies, and lower world petroleum prices.

These new developments are generally to be welcomed by most forest product industries in Latin America and the Caribbean Basin, particularly in the region's petroleum-importing countries. Yet there are winners and losers across a broad spectrum, and welfare changes should not be oversimplified. From the standpoint of the United States, attempts to expand U.S. forest products exports to the region continue to confront the following obstacles:

1. Critical export markets, or potential export markets, are either soft or closed. The economies of two of the region's largest importers, Mexico and Venezuela, depend on the vagaries of world petroleum prices. Trinidad and Tobago, an important importer of U.S. lumber, likewise depends on petroleum earnings. The same can be said of Ecuador and its imports of U.S. paper and paperboard. Cuba, a large importer of forest products, is supplied within the economic orbit of the Soviet Union. Brazil, the largest and wealthiest country of the region, is not a significant buyer of U.S. forest products.

2. In predictions made a few years ago, some U.S. exporters had been optimistic about expanding sales to certain markets, such as U.S. lumber exports to the Caribbean Basin (Robinson 1984). To date those expectations have not been realized. Annual U.S. solid-wood exports to Mexico, the Caribbean Islands, and Central America have remained more or less flat (in current dollars) since 1980 (NFPA 1985). With respect to promoting trade in forest products, the early years of the Caribbean Basin Initiative have been disappointing (Laarman and Muench 1985).

3. Within the region, and also outside of it, the United States can expect increasing export competition from Brazil and Chile. The era of the strong dollar prior to 1985 allowed Brazil to undercut the prices of North American competitors in the European market for market pulp and linerboard (Pearson et al. 1985). Over the last ten years, Brazil's world share of chemical pulp capacity has increased from virtually zero to more than 5% (Slinn 1986). Noticeable quantities of Brazilian pulp are arriving directly in the United States and Canada (Zobel 1987).

4. The arrival of Brazilian pulp in North America may be a red herring. The United States loses jobs and income in primary forestry and pulp manufacture as the direct result of Brazilian imports. But the United States also competes with rapidly increasing Brazilian exports of pulp, paper, and paperboard to the rest of the world. Conceivably, income losses to the U.S. industry are greater for the second reason (loss of export markets, pulp and paper) than for the first one (loss of home markets, pulp). Yet the second impact is less visible, not as easily documented, and much more difficult to measure and evaluate.

5. The ability of Brazil and Chile to compete for exports depends critically on the rapid and dramatic realignments taking place in currency exchange rates. Among pulp traders whose currencies fell against the U.S. dollar, Brazil and Chile headed the list in 1984-85, with depreciations of 65% and almost 50%, respectively (Slinn 1985). More generally, almost all major forest product traders with the United States (both exporters and importers) have seen their currencies depreciate against the dollar. Figure 1 shows exchange rates (ER) as a percentage of the exchange rates that prevailed about ten years ago. In the absence of intervention by central banks to provide a special exchange rate different from the market rate, this has meant that imports from the United States have become increasingly costly, while exports to the United States have become increasingly cheaper. Of course, the issue of cost advantage in a world context is complicated by the realignments of Latin American currencies with currencies other than the U.S. dollar. In a number of countries it is further complicated by hyperinflation and currency instability.

In conclusion, the next few years of forest products trade with Latin America and the Caribbean Basin are likely to be twisted and turned by any number of unpredictable events. Responses to recent political, social, and economic upheavals are still evolving. The very imaginative, and the very naive, think they see clearly tomorrow's round of developments. All others humbly await new adjustments day by day.

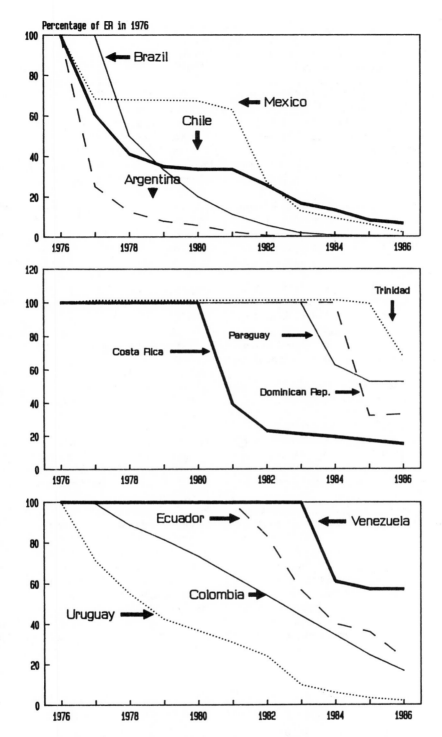

Figure 1. Movement of selected currencies against the U.S. dollar. Source: data published by the International Monetary Fund (1986: spot rate for October).

REFERENCES

Durst, P. B., C. D. Ingram, and J. G. Laarman. 1986. Statistics on forest products trade: Are they believable? *In* G. F. Schreuder (ed.) World trade in forest products 2, pp. 265-273. University of Washington Press, Seattle.

Erfurth, T., and H. Rusche. 1978. The marketing of tropical wood in South America. FAO Forestry Paper 5. FAO, Rome.

FAO. Yearbook of forest products 1973-84. Food and Agriculture Organization of the United Nations, Rome.

———. 1986. Forest products world outlook projections, 1985-2000. FAO Forestry Paper 73. FAO, Rome.

Gonzalez, H. 1985. The Latin American debt crisis: The bailout of the banks. Inter-American Economic Affairs 39:55-70.

Gregersen, H. M., and A. Contreras. 1975. U.S. investment in the forest-based sector in Latin America. Resources for the Future, Johns Hopkins University Press, Baltimore.

IDB. 1985. Economic and social progress in Latin America. Inter-American Development Bank, Washington, D.C.

Iglesias, E. V. 1984. Latin America: Crisis and development options. CEPAL Review 23:7-28.

IMF. 1985. World economic outlook, April 1985. International Monetary Fund, Washington, D.C.

Laarman, J. G. 1986. The economic outlook for forestry in tropical America: A hazardous period for projections. *In* Management of the forests of tropical America: Prospects and technologies. San Juan, Puerto Rico.

Laarman, J. G., and J. Muench, Jr. 1985. The Caribbean Basin Initiative: What does it mean for forestry? *In* Symposium on Sustainable Development of Natural Resources in the Third World. Ohio State University, Columbus.

Lanly, J. P. 1982. Tropical forest resources. FAO Forestry Paper 30. FAO, Rome.

McGaughey, S. E., and H. M. Gregersen. 1983. Forest-based development in Latin America. Inter-American Development Bank, Washington, D.C.

NFPA. 1985. Wood products international trade statistics, vol. 2 (North America, Central America and Caribbean, South America), 1980-1985. National Forest Products Association, Washington, D.C.

Pearson, J., P. Sutton, and H. O'Brian. 1985. World review, Latin America. Pulp and Paper 59(8):59.

Prebisch, R. 1985. International monetary indiscipline and the debt problem. J. Develop. Plan. 16:173-176.

Robinson, S. G. 1984. Marketing opportunities in Latin America. *In* H. E. Dickerhoof, D. Robertson, and J. White (eds.) International forest products trade: Resources and market opportunities, pp. 52-57. Forest Products Research Society, Madison, Wisconsin.

Slinn, R. J. 1985. The case of the confusing pulp market. American Paper Institute, New York, N.Y.

———. 1986. Supply of fiber and its effect on future markets. American Paper Institute, New York, N.Y.

UNIDO. 1983. Tariff and non-tariff measures in the world trade of wood and wood products. UNIDO Sectoral Working Paper Series 6. United Nations Industrial Development Organization, Vienna.

Wisdom, H. W., J. E. Granskog, and K. A. Blatner. 1986. Caribbean markets for U.S. wood products. USDA Forest Service Res. Pap. SO-225. Southern Forest Experiment Station, New Orleans.

Yoho, J. G., and C. M. Gallegos. 1984. Timber supply and market opportunities in Latin America. *In* H. E. Dickerhoof, D. Robertson, and J. White (eds.) International forest products trade: Resources and market opportunities, pp. 45-51. Forest Products Research Society, Madison, Wisconsin.

Zobel, B. J. 1987. Forestry potential in South America. Society of American Foresters, Appalachian Section Meeting, Greenville, South Carolina.

Chapter 2

The Latin American Southern Cone's Role as a New Wood Supplying Region

MARCIO A. R. NAHUZ

Despite retaining one-quarter of the world's timber growing stock in operable forests, South America has always contributed less than one-tenth of the production of any known commodity, from roundwood to paper and paperboard, except for fuelwood and charcoal. Within the region, Argentina, Brazil, and Chile stand out in forest productivity, production either actual or potential, consumption, and exports of timbers and wood-based products. Growing out of the traditional pattern of low-productivity forests and industries, and minor contributions to the world trade in forest products, these countries are attaining a position of competitiveness by producing and exporting significant volumes of high-quality coniferous and broadleaf timbers to the main consuming markets.

This paper reviews the forest potentialities and the performance of the forest sector in these countries, discusses the main factors affecting the established patterns, and analyzes the trends and influence of these new wood-supplying regions. Four topics are covered: (1) South America as a producer and exporter of forest products, (2) a description of the Southern Cone, (3) the three major producers and consumers, and (4) major trading patterns.

SOUTH AMERICA AS PRODUCER AND EXPORTER

The forest area of the world is estimated in 1987 at about 4,370 million hectares, covering about one-third of the world's land area (based on FAO 1985, updated with annual rates of forest loss and renewal). Of these, 900 million hectares—20% of the world's total—are in South America, an area equivalent to about 96% of the total area of the United States (this includes Argentina, Bolivia, Brazil, Chile, Colombia, Ecuador, French Guiana, Guyana, Paraguay, Peru, Suriname, Uruguay, and Venezuela). About half of the South American forests are considered operable or productive— that is, available and allowing for the production of industrial roundwood (sawlogs and veneer logs, pulpwood, and other industrial wood).

Figure 1 shows the percentage contribution of the region in the production of various forest products in terms of volume of both the coniferous and broadleaf material generated in selected years. With few exceptions, the South American contribution, in any year and in any product, has been less than 10% of the world's total production. The exceptions mentioned are the fuelwood and charcoal produced in the region, which represented up to 17% in 1971, then declined to a constant 13% in the last decade. However, this is still an index generally not too desirable in terms of industrial development.

The general trends detected in Figure 1 indicate a rising production of industrial roundwood, particularly for pulpwood and particles. Production growth is also seen in sawnwood, wood-based panels, woodpulp, and paper and paperboard. The combination of a rising trend in the production of these wood-based commodities with the more or less stable production of total roundwood, as

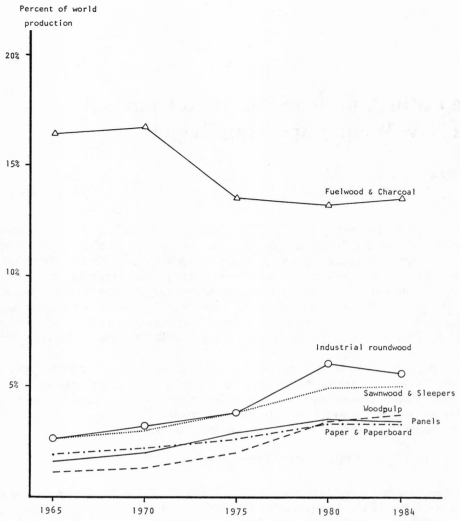

Figure 1. Percentage contribution of South America to production of forest commodities, 1965-84.

well as the decline of fuelwood and charcoal production, suggests a steady growth in industrial development in the region.

This increase in the South American contribution to world production is explained not by the eventual decline of the relative contribution of countries outside the region, since this has been restricted to a minimum in the last decade (FAO 1986), but in part by an increasing number of forest-based industries being established in South America (or the expansion of those already operating there) and in part by a significant growth in forest and industrial productivity.

In the international market of forest products, South America has not yet played an important role. The relative share of the region in world exports has remained at the lowest levels in practically all commodities. This less-than-desirable performance is sketched in Figure 2, which shows a minimum of exports in roundwood form. A better picture is seen in the exports of industrialized products with a much higher market price, such as woodpulp, wood-based panels, paper and paperboard, and even sawnwood. Exports of the first three commodities have increased consistently, both in relative participation and actual volumes.

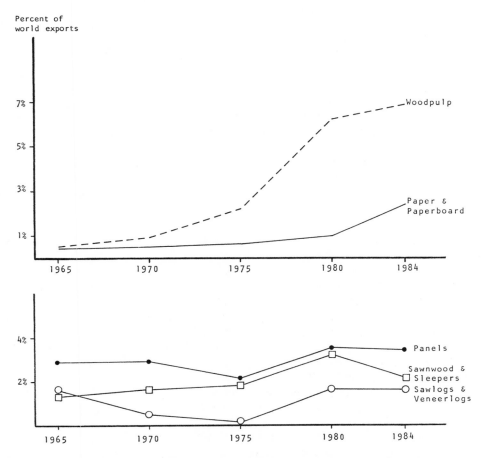

Figure 2. Percentage contribution of South America to world exports of forest products, 1965-84.

In terms of value, the situation is similar to that of export volumes. The South American share in the total value of world exports of forest products was under 1.5% up to 1975, and only in 1980 did it rise to 2.7%, reaching 3.2% of the world total in 1984, with a little more than U.S.$1,600 million.

In summary, the performance of South America in production and export of forest products has been negligible in the context of world production and trade, particularly in view of the region's immense forest resources and production potential. However, the region is changing. The fast pace of the expanding demand, mostly domestic, requires increasingly larger volumes of industrial wood, as well as wood for energy generation. This in turn demands better productivity from both forests and industries, as well as larger areas of plantations.

THE SOUTHERN CONE OF SOUTH AMERICA

Although geographically the Southern Cone of South America consists of Argentina, Chile, and Uruguay, it would be unrealistic to speak of the forest resources of the region without including Brazil. The four countries are discussed separately below.

Uruguay

Among the countries mentioned, Uruguay has the smallest area—about 176,000 square kilometers, with a total forested area of some 615,000 hectares. Of these, 70% are native subtropical and temperate forests situated mostly along the rivers, and they are under protection. These forests do not represent an exploitable resource. The forest plantations amount to about 170,000 hectares, consisting of eucalypts (70%), pines (17%), and willows and others (13%) (Kalas 1982).

In 1984, Uruguay produced 3 million m^3 of roundwood (93% of this as fuelwood and charcoal). The country is a net importer of sawnwood, wood-based panels, woodpulp, paper, and paperboard (FAO 1986).

Uruguay is currently completing a survey of the country's forest resources with the help of satellite imagery, as a first step in the expansion of its forest plantations. The country expects to substitute a substantial part of the imported fossil fuel used in the industry with locally grown biomass.

Argentina

Argentina has a land area of about 2.8 million square kilometers, with a forest cover estimated at 44.7 million hectares, of which 98% is native forests.

Native Forests. The native forests of Argentina range from rainy subtropical forests of the Northeast and Northwest—"Selva misionera" and "Selva tucumano-boliviana"—to the subantarctic forests of the South and Southwest. These are typical forest formations, yielding timber for both industrial uses and energy, and make up about 13.6% of the native forests. In addition, large areas of the country are covered by the mostly hot and dry heavily wooded savannas, regionally known as *parques (parque chaqueño, parque pampeano-puntano, parque mesopotámico)* yielding mostly fuelwood but also good and durable woods. These formations account for a little over 68% of the native forests of the country (Tinto 1986).

Practically all the native forest is broadleaf and none of it is under management. Estimates of the growing stock are low, ranging from 5 to 20 m^3/ha in the more poorly stocked areas to about 110 m^3/ha in the richer formations.

Among the favored native species in Argentina (Huek 1972, Tinto 1979), the most important are urunday (*Astronium balansae*), lapacho (*Tabebuia* spp.), ibirapuitá (*Peltophorum dubium*), anchico colorado (*Piptadenia rigida*), and cedro (*Cedrela* spp.) in the subtropical and riverine forests; and algarrobo (*Prosopis* spp.), quebracho (*Schinopsis* spp. and *Aspidosperma* spp.), and palo santo (*Bulnesia sarmientoi*) in the dry, open forest and *parques.*

Some native conifers also occur but are relatively scarce. These are Parana pine or Pino de Misiones (*Araucaria angustifolia*), ciprés (*Libocedrus chilensis*), mañiu (*Podocarpus* spp.), and pehuén (*Araucaria araucana*).

In the humid valleys and in different latitudes and altitudes, the mixed forests of conifers and broadleaf species include roble, lenga, coigue, and rauli (*Nothofagus* spp.).

Plantations. The forest plantations in Argentina were first established as protection forests, in isolated stands or as shelterbelts. More recently, the Argentinian government, in an attempt to increase productive forests, established the National Reforestation Plan, which provides fiscal incentives in selected regions to approved projects of reforestation using fast-growing species. With this plan, Argentina intended to reforest an area of 360,000 hectares from 1978 to 1982. About half of this area was effectively planted (IFONA 1982).

The statistics on reforestation in Argentina are not up to date; the most recent information available is for 1983 (Table 1). The forest plantations existing in 1983 were estimated at about 730,000 hectares; the comparison of this area with that listed for 1979 shows a 10% increase in the planted area in four years.

Most of the plantations (43% of the 1983 area) consist of conifers. These include pines (*Pinus radiata, P. caribaea, P. elliottii, P. taeda,* and others), Parana pine, Douglas-fir (*Pseudotsuga menziesii*), and cypress (*Cupressus* spp.). The eucalypt plantations account for 28% of the reforested

Table 1. Area of forest plantations in Argentina in selected years.

Plantation	Area (1,000 hectares)		
	1979	1980	1983
Conifers	280.1	321.1	316.8
Eucalypts	174.8	192.8	203.3
Poplars and willows	162.5	184.9	191.4
Other	44.5	18.5	20.6
Total	661.9	717.3	732.1

Sources: Tinto (1986), Argentina (1984).

area; these include *Eucalyptus citriodora*, *E. camaldulensis*, *E. viminalis*, *E. saligna*, *E. grandis*, and *E. tereticornis*. Poplars and willows (*Populus* and *Salix* spp.) are the third most important group used for reforestation; they represent 26% of the plantations existing in 1983.

The forest plantations of Argentina are concentrated in the northern provinces, close to the main consuming centers, and in regions where the best growth rates can be attained. Misiones, Corrientes, Buenos Aires, and the Parana Delta retain two-thirds of the plantations established up to 1983. These are mostly young forests: nearly 70% of the conifer stands are fourteen years old or younger, as are 84% of the eucalypts (Tinto 1986). This age distribution means that in the next three or four years a considerable area of both softwoods and hardwoods will come into production.

By South American standards, the growth rates of the Argentinian plantations are relatively low. Typical mean annual increments in volume of sawmill stock (Tinto 1986) are: eucalypts, 16 m^3/ha per year; conifers, 12 m^3/ha per year; and poplars and willows, 9 m^3/ha per year. These growth rates, applied to the existing plantations, indicate a potential annual crop of 8 million m^3 (5 million m^3 for sawnwood and 3 million m^3 for other uses, including woodpulp, panels, poles, and energy).

Wood Processing Industries. The wood processing industries of Argentina are mostly of small capacity and in recent years have been operating below capacity. However, with the availability of locally grown raw material expected in the near future, added to a growing consumption of all industrialized wood-based products, it is likely that the forest-based industries will soon be operating at full capacity.

The Argentinian Institute of Forestry statistics show that in 1983 there were 2,000 sawmills operating in the country. Despite an installed capacity of 1.7 million m^3, the production of this sector reached only 1.1 million m^3, two-thirds of capacity. Similar data are seen in Table 2 for other types of wood industries, showing the installed capacities and the 1983 production in Argentina.

Table 2. Wood processing industries in Argentina.

Type of Industry	No. of Establishments	Installed Capacity (1,000 m^3 or m.t.)	Production 1983 (1,000 m^3 or m.t.)
Sawnwood	2,000	1,680	1,120
Wood-based panels:			
Veneer	20	41	23
Plywood	23	94	48
Fiberboard	2	121	94
Particleboard	9	466	241
Woodpulp	21	800	630
Paper and paperboard	78	1,200	900

Sources: IFONA (1982, 1985).

Brazil

Brazil is the largest country in South America, with a land area of 8.5 million square kilometers, of which 65% is still covered with forests (FAO 1985). Because of its width (34°W to 74°W) and length (5°N to 34°S), Brazil covers numerous forest regions, from the humid equatorial zone to the cold temperate regions of the South.

Native Forests. These are basically divided into three types: (1) Amazon, (2) Atlantic, and (3) mixed forests of araucaria.

The Amazon forest is situated in the northern region of Brazil, covering an area of 280 million hectares (Figure 3). It includes some 6.5 million of the periodically flooded "várzea" forests, along the main rivers, and 27.4 million hectares of the "terra firme" forests, more heavily stocked but of more difficult access.

Traditionally the "várzea" forests have been exploited and yielded the lighter-colored and lightweight timbers such as *Virola*, *Ceiba*, and *Hevea*, widely used for veneer and plywood production. The standing stock in these forests averages 90 m^3/ha, of which one-third consists of commercial species.

The "terra firme," or highland, forests are now under exploitation, yielding increasing volumes of both fuelwood and industrial roundwood. These forests generate the heavier and stronger timbers used in construction, such as cupiuba (*Goupia glabra*), jatoba (*Hymenaea* spp.), ipê (*Tabebuia serratifolia*), and more than one hundred other species.

Other favored timbers in these forests that are widely accepted in the international markets are mahogany (*Swietenia macrophylla*), andiroba (*Carapa guianensis*), Spanish cedar (*Cedrela* spp.), and satinwood (*Euxylophora paraensis*).

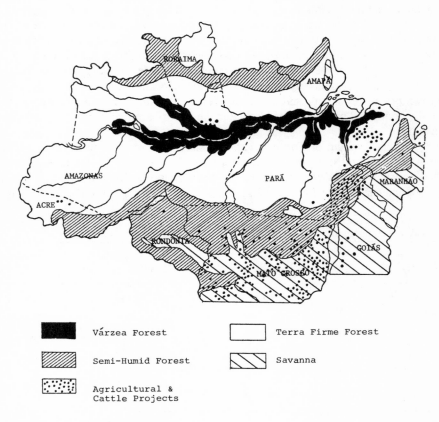

■ Várzea Forest	□ Terra Firme Forest
▨ Semi-Humid Forest	⬡ Savanna
⠿ Agricultural & Cattle Projects	

Figure 3. Brazilian Amazon Region forest formations.

The standing stock in the "terra firme" forests varies between 100 and 300 m^3/ha, with an average of 180 m^3/ha of trees over 25 cm in diameter. Of this volume about 25%, or 45 m^3/ha, is destined for the domestic markets and some 20 m^3/ha consist of species traditionally exported.

The Atlantic forest covers a strip of variable width along the Atlantic coast from 6°S to 30°S, with a total area not exceeding 3 million hectares. This forest contains prime species such as rosewood (*Dalbergia nigra*), caviuna (*Machaerium scleroxylon*), vinhático (*Platymenia reticulata*), and others. Large areas of these forests are now under protection.

Mixed forests of araucaria occur in the highlands of southern Brazil, in the more temperate regions. Parana pine is the dominant species, mixed with *Podocarpus* spp., imbuia (*Ocotea porosa*), and other *Nectandra* and *Ocotea* species. These forests originally occupied some 8 million hectares, but heavy exploitation over the years has drastically reduced the area so that by 1950 the forests amounted to 2.5 million hectares. At present the native araucaria forest does not exceed 270,000 hectares (IBDF 1983). Such forests used to stock 270 m^3/ha of prime Parana pine. No such forest will ever exist again.

Plantations. The first planted forests were established in Brazil with eucalyptus at the beginning of this century. The main objective was fuelwood production. The planting activity increased at a modest pace, including the introduction of exotic conifers, to be used in the production of pulpwood.

In 1966 the implementation of a policy of fiscal incentives for reforestation encouraged this activity to such a degree that in 1976 and 1979 plantations reached a peak of about 470,000 hectares per year. Table 3 shows the reforested areas in Brazil, in selected years.

By 1984 the total planted area in Brazil amounted to 5.6 million hectares—an average of 310,000 hectares per year (IBDF 1985). More than half of the area (3 million ha) is planted with *Eucalyptus* species, mainly *E. saligna*, *E. urophylla*, *E. citriodora*, *E. grandis*, and others. These are grown for the production of charcoal used to fire the Brazilian steel mills, for pulpwood, poles, and as raw material for fiberboard, particleboard, and, more recently, sawnwood. Eucalypt plantations are concentrated in the Southeast and Southwest.

The mean annual increments of eucalyptus plantations vary according to soil, species, and silvicultural and management techniques; values from 5 to 35 m^3/ha per year are easily found. Appropriate species selection, seed improvement, clonal propagation, effective pest control, and better management techniques are helping to improve yields. Experimental stands of eucalyptus have reached yields of 75 m^3/ha per year. Pine plantations cover 1.8 million hectares, the main species being *Pinus elliottii*, *P. taeda*, *P. caribaea*, and *P. oocarpa*. These are grown for pulp, particleboard, and sawnwood. Mean annual increments of 8 to 24 m^3/ha per year have been reported

Table 3. Area of forest plantations in Brazil in selected years, with fiscal incentives for 1967-84.

Year	Area (1,000 hectares)			
	Conifers	Eucalypts	Other	Total
1967	20	14	1	35
1970	132	84	6	222
1974	91	188	45	324
1976	112	262	95	469
1979	119	282	72	473
1981	117	230	71	418
1984	96	162	42	300
1967-84	1,812	3,036	743	5,591
	32.4%	54.3%	13.3%	100%

Source: IBDF (1985).

(Ponce 1982). These plantations are concentrated in the South, but the tropical pines are planted mostly in the Southeast.

Wood Processing Industries. The most recent statistics on the wood processing industries of Brazil indicate that there are almost 14,000 sawmills in the country. This number may fluctuate according to demand, availability of logs, and market prices of sawnwood (Table 4).

Table 4. Wood processing industries in Brazil.

Type of Industry	No. of Establishments	Production (1,000 m^3 or m.t.)
Sawnwood	13,770	15,852 (1984)
Wood-based panels	660	2,523 (1984)
Woodpulp	82	4,485 (1985)
Paper and paperboard	150	4,021 (1985)

Sources: ANFPC (1985), IBDF (1985), UFRRJ (1984)

Capacities, production, degree of mechanization, and type of equipment used vary widely within the country. A typical mill operating in the Amazon would work a bandsaw headrig, one or two single circular edgers, and one or two crosscut saws, with an output of 300 to 600 m^3 per month (Ponce 1982).

Up to the mid-1970s, most sawmills were concentrated in the South, close to the araucaria forests. More recently, there has been a distinct move toward the more promising forests of the North, where a number of large industries were established to produce sawnwood and sleepers.

The wood-based panel industry in Brazil produces fiberboard (from eucalypts), particleboard from pine and mixed species, and plywood and veneers both from Parana pine and other native species. A number of plywood and veneer plants are also moving to the North or West, because of the shortage of veneer logs in the South and East.

The pulp and paper industry is concentrated in the Southeast and South, and uses eucalypts for short fibers and pine and araucaria for long fibers. The 1985 production reached nearly 4.5 million metric tons of pulp and 4 million of paper and paperboard (ANFPC 1986).

Chile

Chile has a total area of 757,000 square kilometers, with a forest area of 9 million hectares—that is, 12% of the country (FAO 1985). Stretched between parallels 17°S and 55°S, Chile is climatically divided into four regions; its forest resources are concentrated in the central, southern, and austral regions.

Native Forests. The native forests of Chile are of eleven main types. Of these, the most important are:

1. Roble, rauli, and coigue forest (36°30′S to 40°30′S): of great economic value, yielding wood for joinery, framework, paneling, and veneer. The main species are *Nothofagus obliqua, N. procera,* and *N. dombeyi.*

2. Lenga forest (36°50′S to 56°S): mainly *Nothofagus pumilio*; can be pure or associated with araucaria and *N. dombeyi.*

3. Araucaria forest (37°S to 40°48′S): associated with lenga or lenga and coigue.

4. Evergreen forest (40°30′S to 47°S): of variable composition.

5. Alerce forest (39°50′S to 43°30′S): mainly *Fitzroya cupressoides,* exploited because of its durability.

Of the native forests of the country, 7.6 million hectares are considered commercial forests (i.e., with more than 30 m^3/ha of trees with diameters of 25 cm and above), accessibility not considered. The standing stock in these forests is estimated at 915 million m^3 (CORFO-INFOR 1986).

Plantations. The forest plantations in Chile cover 1.2 million hectares, consisting mainly of radiata pine (*Pinus radiata*). Table 5 summarizes the species planted to the end of 1985. Annual rates of reforestation average 83,210 ha per year (1975-85); in 1985 the plantations with radiata pine amounted to 84% of the total area. Planting is carried out by both the state organization CONAF (Corporacion Nacional Forestal) and private companies; these were responsible for 72% of the area planted between 1975 and 1985 (CORFO-INFOR 1986).

Table 5. Areas of forest plantations in Chile, to the end of 1985.

Species	Area (1,000 ha)	%
Radiata pine	1,040.25	87.5
Eucalypts	51.17	4.3
Atriplex spp.*	24.10	2.0
Prosopis spp.**	23.74	2.0
Douglas-fir	10.54	0.9
Poplars	3.23	0.3
Other	35.60	3.0
Total	1,188.63	100.0

Source: CORFO-INFOR (1986).
*For forage.
**Tamarugo and Algarrobo.

Radiata pine found in Chile excellent conditions for its development. The mean annual increments in volume, estimated on a nationwide basis, vary as seen in Table 6. At these rates of growth, estimates are that the forest plantations of radiata pine in Chile stocked at end of 1985 had a potential of 100 million m^3. Areas and volumes according to age bracket are also seen in Table 6.

Table 6. Areas, volumes, and mean annual increments in volume of radiata pine plantations in Chile.

Age Bracket (years)	Area (1,000 ha)	Volume (1,000 m^3)	Mean Annual Increment in Volume (m^3/ha)
1-5	351.5	--	--
6-10	325.2	--	--
11-15	195.2	30,969	10.6-14
16-20	96.1	29,908	15.6-19.5
21-25	34.7	16,145	18.6-22.1
26-30	25.3	14,895	19.6-22.7
31+	12.3	8,540	--
Total	1,040.3	100,457	

Source: CORFO-INFOR (1986).

Wood Processing Industries. Sawnwood production consumes most of the roundwood produced in Chile—a 44% share compared with 39% for woodpulp—in the period 1974-85 (CORFO-INFOR 1986).

The main species used is radiata pine; it amounted to 1.87 million m^3 of the 2.19 million m^3 produced in 1985 (Table 7). The production of radiata pine sawnwood has increased substantially since 1974, from an average of 470,000 m^3 per year to the present levels of 1.8 million m^3 per year. Other species used in sawnwood production are mainly *Nothofagus*.

Table 7. Wood processing industries in Chile.

Type of Industry	Installed Capacity 1985 (1,000 m.t.)	Production 1985 (1,000 m^3)
Sawnwood		2,191*
Wood-based panels		230
Woodpulp	876**	837
Paper and paperboard	432†	370

Sources: CORFO-INFOR (1986), CICEPLA (1986).
*Radiata pine, 85.4% = 1,871,000 m^3.
**Projected 1987: 1,087,000 m.t.
†Projected 1987: 605,000 m.t.

Chile also produces fiberboard, particleboard, plywood, and veneer. The 1985 production of these wood-based panels was 230,000 m^3. Most of the raw material used is also radiata pine—71% of the total volume of wood consumed by this sector in 1985. Eucalypts and native species account for the balance.

The woodpulp industry in Chile had an installed production capacity of 876,000 m.t. in 1985. In that year, its production reached 837,000 m.t., 95% of the total capacity, which is expected to increase to over 1 million m.t. in 1987. Some 97% of the wood volume consumed in this sector in 1985 was radiata pine.

A similar trend is seen in the paper and paperboard industry.

PRODUCTION AND CONSUMPTION: ARGENTINA, BRAZIL, AND CHILE

The area of native forests in South America is decreasing at the rate of 3.1 million hectares per year, one half of it in Brazil (FAO 1985). This is mostly the result of agricultural and cattle raising projects, the construction of hydroelectric plants, migration and resettling, and shifting agriculture. This deforestation is concentrated in the Amazon Region (Figure 4), but is being monitored closely and has started to decrease in intensity.

At the same time, reforestation is taking place at a fast pace: Argentina: 170,000 ha/year (projection, 1987-94); Brazil: 310,000 ha/year (average 1967-84); Chile: 83,000 ha/year (average 1975-85). If half of these areas are effectively planted, this will amount to an additional 3.7 million hectares of productive forests in the Southern Cone by the year 2000. At a conservative rate of growth, this represents a potential of 366 million m^3 of both conifers and eucalypts, mostly industrial roundwood.

The production of industrial roundwood in Brazil, Chile, and Argentina increased in the period 1975-84, by 88%, 73%, and 37%, respectively. Apparent consumption, however, also increased—by 87%, 56%, and 37%, respectively—indicating the net availability of saw material for export,

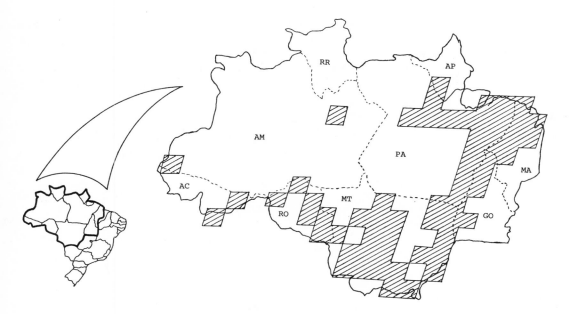

Figure 4. Timber production areas in the Amazon Region, with critical areas of deforestation. Source: IBDF (1985).

either as roundwood (from Chile) or as processed wood products. Unfortunately, there are no reliable long-term projections available of the production of roundwood on a country-by-country basis.

Table 8 shows the production, exports, and apparent consumption of sawnwood and sleepers for Brazil, Chile, and Argentina from 1975 to 1984. The actual growth rates are indicated, together

Table 8. Production, exports, and consumption of sawnwood and sleepers in the Southern Cone (1,000 m^3).

Country	1975	1980	1984	% Growth 1975-85	1984-2000 FAO*	Chase**
Brazil						
Production	10,129	14,881	15,852	56	135	81
Exports	500	809	535	7		
Apparent consumption	9,706	14,534	15,656	61	115	90
Chile						
Production	1,320	2,186	2,001	52	27	31
Exports	261	1,298	886	39		
Apparent consumption	1,059	888	1,115	5	–10	–6
Argentina						
Production	480	883	1,237	158	37	18
Exports	--	1	--	--		
Apparent consumption	1,043	1,525	1,558	49	6	6

Sources: FAO (1986a, b).
*Based on the FAO Compendium of Macro-Economic Indicators.
**Chase Econometrics 1986 Long Term International Forecast.

with the long-term projection carried out by FAO and Chase. Table 8 indicates the potential availability of sawnwood for the export market, since growth rates of production are higher than those of consumption.

In fact, rough calculations indicate that by the year 2000 there will be a sawnwood surplus production of 3.6 million m^3 in Brazil, 1.5 million m^3 in Chile, and 46,000 m^3 in Argentina, which will be available for export.

Table 9. Production, exports, and consumption of wood-based panels in the Southern Cone (1,000 m^3).

Country	1975	1980	1984	% Growth 1975-84	% Growth 1984-2000 FAO*	% Growth 1984-2000 Chase**
Brazil						
Production	1,725	2,482	2,523	46	216	148
Exports	185	324	453	45		
Apparent consumption	1,546	2,246	2,123	37	365	294
Chile						
Production	44	115	183	316	58	68
Exports	--	32	50	--		
Apparent consumption	44	83	133	202	83	102
Argentina						
Production	331	414	390	18	149	108
Exports	8	22	23	187		
Apparent consumption	332	410	378	14	113	111

Sources: FAO (1986a, b).
*See Table 8.
**See Table 8.

Table 10. Production, exports, and consumption of total woodpulp in the Southern Cone (1,000 m.t.).

Country	1975	1980	1984	% Growth 1975-84	% Growth 1984-2000 FAO*	% Growth 1984-2000 Chase**
Brazil						
Production	1,208	3,047	3,433†	184		
Exports	153	890	976	538		
Apparent consumption	1,156	2,217	2,478	114	103	65
Chile						
Production	452	763	839	86		
Exports	174	416	493	183		
Consumption	285	347	346	21	21	33
Argentina						
Production	259	309	561	117		
Exports	--	--	--	--		
Consumption	372	179	630	69	33	30

Sources: FAO (1986 a, b).
*See Table 8.
**See Table 8.
†Brazilian production increased 1.5% in 1985.

The same trends do not apply to wood-based panels (Table 9). In both Brazil and Chile, consumption is expected to grow at a faster pace than production. Argentina may become a net supplier of panels. Table 10 summarizes the performance for woodpulp production and consumption, but the estimated rates of production growth are not available. Table 11 lists the information for paper and paperboard.

Table 11. Production, exports, and consumption of total paper and paperboard in the Southern Cone (1,000 m.t.).

Country	1975	1980	1984	% Growth 1975-84	1984-2000 FAO*	Chase**
Brazil						
Production	1,688	3,361	3,768†	123	109	66
Exports	7	198	703	9,943		
Apparent consumption	1,898	3,424	3,281	73	186	146
Chile						
Production	235	356	375	60	47	52
Exports	107	87	136	27		
Apparent consumption	135	323	297	120	63	78
Argentina						
Production	650	713	946	45	49	38
Exports	12	15	12	--		
Apparent consumption	816	947	1,031	26	46	45

Sources: FAO (1986 a, b).
*See Table 8.
**See Table 8.
† Brazilian paper and paperboard production in 1985 increased by 7.5% over 1984.

TRADE PATTERNS

The major trading partners of Brazil, Chile, and Argentina are undoubtedly the developed countries of North America, Western Europe, and the Far East. Based on 1984, 1985, and part of 1986 export statistics, the United States is the most important consumer of Brazilian forest products. The direction of trade of forest products from Brazil and Chile is seen in Table 12.

Brazilian sawnwood, fiberboard, and paper are exported mainly to the United States. Woodpulp finds its markets mainly in Western Europe and Japan.

Most of the sawnwood exported is *Swietenia* mahogany. Official statistics show that one-third of the total exported in 1985 was mahogany sawnwood; other sources estimate the mahogany share at 50%, the difference being exported under trade names other than mahogany.

Other species are finding their way into the international markets. At present there is a drive toward the introduction of lesser known tropical timbers into the export markets. With a foreseeable scarcity of mahogany logs, this is likely to succeed.

There is also a drive in Brazil toward exports of tropical logs, from areas to be flooded due to the construction of hydroelectric plants. Japan, China, and Korea have shown much interest, and two experimental log shipments have already been made to China.

The wood produced in plantations is consumed domestically, either as industrial roundwood or as fuelwood and charcoal. There is no potential yet for exporting roundwood.

Table 12. Direction of trade of main forest products from Brazil and Chile (1984-86).

Producer	Commodity	Consumer	% Total Exports
Brazil	Sawnwood	United States	40
		Great Britain	22
	Plywood	Great Britain	30
		Latin America	28
		Arab countries	15
	Fiberboard	United States	53
	Woodpulp	Belgium/Luxemburg	27
		Japan	20
		United States	14
	Paper and paperboard	United States	18
		Africa	16
		Latin America	14
Chile	Logs	Korea	46
		Japan	30
		China	22
	Sawnwood	Arab countries	44
		Latin America	27
		Japan	15
	Panels	United States	49
		Korea	19
	Woodpulp	Latin America	13
		West Germany	11
		China	10
	Paper and paperboard	Latin America	47

Sources: FAO (1986), CORFO-INFOR (1986), and IBDF (1986).

Chile has been exporting logs of radiata pine since 1975, but starting in 1979 both production and exports rose appreciably. At present, Chilean logs compete in the Far East with conifer logs from the U.S. Northwest Coast.

Log prices in U.S. dollars are very competitive. Nominal values have decreased from U.S.\$48/m^3 f.o.b. in 1981 to U.S.\$31/m^3 f.o.b. in 1985 (CORFO-INFOR 1986). Freight costs to the Far East ports range from U.S.\$15 to \$25/m^3.

Chilean exports in the first five months of 1986 have registered increases: radiata pine sawnwood, 30% over the same period in 1985; radiata pine logs, 4%. The country is now beginning to export pulpwood chips: 95,000 m^3 were shipped to Finland in May 1986.

Argentina strives to reduce its imports of wood and most forest products by expanding its area of productive forest plantations. It will be a long while yet, but the country will probably become a net supplier of panels and pulp in the future.

It may be that the Southern Cone of South America (including Brazil) is not yet a massive wood-supplying region. But those countries have enormous areas suitable for forest plantations, enough sunlight hours, and good climates for forest development. Furthermore, effective silvicultural techniques are being developed and applied, and the results are already in view.

In addition, tropical countries such as Brazil are being urged, both externally and internally, to develop a more rational approach to the utilization of their immense native forest resources, curbing deforestation and increasing the usable volume of wood per unit area.

A growing demand for all types of forest products seems to be more than ever the driving force behind forest development. Whether those countries of South America will effectively respond to it remains to be seen.

REFERENCES

ANFPC. 1986. Relatório estatístico 1985. Associação Nacional de Fabricantes de Papel e Celulose, São Paulo.

Argentina. 1984. Anuário estadístico de la Republica Argentina. Ministero de Economia, Buenos Aires.

CICEPLA. 1986. Estadisticas anuales 1985. Confederatión Industrial de la Celulosa y el Papel Latino americano.

CORFO-INFOR. 1986. Estadisticas forestales 1985. Corporacion de Fomento de la Producción y Instituto Forestal, Santiago.

FAO. 1976. Yearbook of forest products 1963-74. Food and Agriculture Organization of the United Nations, Rome.

——. 1985. Forest resources 1980.

——. 1986a. Yearbook of forest products 1973-84. Rome.

——. 1986b. Forest products world outlook projections, 1985-2000. FAO Forestry Paper 73. FAO, Rome.

Hueck, K. 1972. As Florestas da América do Sul: Ecologia, composição e importância econômica. São Paulo.

IBDF. 1983. Inventário florestal nacional. Sintese dos Resultados. Instituto Brasileiro de Desenvolvimento Florestal, Brasília.

——. 1985. O Setor Florestal Brasileiro 79/85. IBDF, Brasília.

——. 1986. Infoc Madeireiro. Departamento de Industrializaçao e Comercializaçao. IBDF, Brasília.

IFONA. 1982. Informe para la Conferencia Técnica Regional del BID. *In* The forest sector: Country reports. Inter-American Development Bank, Washington, D.C.

——. 1985. Anuário de estadistica forestal 1983. Ministerio de Economia, Secretaria de Agricultura, Ganaderia y Pesca, Buenos Aires.

Kalas, P. 1982. Recursos energeticos nuevos y renovables Uruguay. UNIDO, Vienna. 104 p.

Nahuz, M. A. R. 1985. Recursos florestais Brasileiros. Associação Brasileira de Preservadores de Madeira. Bol. 36. São Paulo.

Ponce, R. H. 1982. Review of the wood and wood products industry in selected countries of Latin America. UNIDO. ID/WG. 380/1.

Tinto, J. C. 1979. Utilización de los recursos forestales argentinos. Instituto Forestal Nacional, Argentina.

——. 1986. Situacion forestal argentina. Su insercion enel concierto mundial. Secretaria de Ciencia y tecnica, Subsecretaria de Coordenacion y Planificacion, Argentina. Buenos Aires.

UFRRJ. 1984. Análise do setor industrial florestal. Universidade Federal Rural do Rio de Janeiro.

UNIDO. 1983. Wood resources and their use as raw material. Sectoral Studies Series 3. IS. 399. United Nations Industrial Development Organization, Vienna.

Chapter 3

Promotional Activities in Latin America: The Demonstration House

DAVID L. ROGOWAY

BACKGROUND

I was pleased to be invited to speak on Latin America at this symposium, because that is where our international efforts outside of Europe began, and it continues to be a region of great interest for the American Plywood Association. Over the past five years, trade missions and wood housing demonstration projects have taken me to many of the Latin American countries. There have been delays and disappointments. But every trip has reinforced my belief that the region presents strong opportunities for wood product exports.

When we look at our American neighbors to the south, it's easy to lump all the countries together as one market. But they are not one market, and that's what makes trade with Latin America so challenging.

First, consider the diversity of the region. There are the hot, tropical islands of the Caribbean. Before we can sell wood housing in the Dominican Republic, Jamaica, or Barbados, we have to convince housing officials and the people there that treated U.S. wood materials can resist damage from termites and high moisture, and that properly designed wood construction will weather the violent hurricanes so common to the region.

In Central America, housing officials have the same concerns about durability, and they have to build for earthquake resistance. In Guatemala, people are still living in temporary shelters erected after the devastating 1974 earthquake. These shelters, no better than livestock pens, are dramatic testimony to the need for durable, permanent housing.

People in El Salvador are particularly conscious of earthquake-resistant construction after a severe quake in 1986. A demonstration house we recently constructed in San Salvador is being used as the administrative office of a field hospital for children. The original concrete hospital was heavily damaged in the quake. This house is the only permanent structure in a community of tents.

In South America, conditions range from the coastal, almost desert climate of Lima, Peru, to subfreezing temperatures and 100 mile-an-hour winds in Tierra del Fuego in southern Argentina.

The requirements for housing are diverse, but the need is universal. All told, Latin America faces a housing shortage of five to six million units per year. The very diversity of this region is what makes wood an ideal solution. Wood construction is adaptable to warm and cold climates, and can be designed to meet special requirements for resistance to termites, moisture, wind, earthquakes, and snow. In U.S. wood products we have a solution for the housing needs of this diverse region. But how do we go about selling that solution?

Frequently the process starts with a trade mission made up of government and industry repre-sentatives. The trade mission gives us a chance to view and evaluate firsthand the potential market for U.S. wood exports. We consider the nation's housing needs, tariff barriers, code restrictions, and local construction methods and traditions. We also establish contacts in the country. These include designers, code officials, members of the distribution trade, and government officials. It is important that we have contacts in both the public and private sectors.

Once we have explored the needs and conditions for trade, we develop a strategy to capitalize on both the immediate and long-term opportunities. This often means working to reduce tariffs, break nontariff barriers, and beat down code restrictions.

THE DEMONSTRATION HOUSE

One of the best tools we've found to present wood construction concepts to Latin Americans is the demonstration house. Since 1982, we have constructed demonstration houses in Venezuela, Chile, Peru, Ecuador, El Salvador, Guatemala, Barbados, the Dominican Republic, and Jamaica. The houses are usually constructed by local crews and a U.S. supervisor.

In working with local builders, we lay the foundation for continued interest and additional construction projects. For example, a U.S. supervisor worked with a large construction company in Guatemala City to complete two demonstration units last fall. The local builder was so impressed with the speed and ease of wood construction that he is now pursuing plans for a 114-unit housing development.

We also maintain the momentum from our demonstration projects with wood construction seminars directed toward builders, designers, and code officials. Seminars have either been given or are scheduled in every country where we've built demonstration houses.

Constructing these houses and developing follow-up promotion is no small task. Anyone involved in programs like this sooner or later is confronted with the question: Given the mounting debts and pessimistic economic forecasts in Latin America, is it worth our time to be there, building demonstration houses and promoting our products? Many reasons are advanced why we shouldn't be there.

First, there's the challenge of dealing with governments fraught with internal politics. In some cases, changing governments and revolving policies make it tough to get to the right person. Frequently, as in Brazil, internal government dealings can have a profound effect on the economics and trade policies of a nation.

But while government wheeling and dealing can deter our efforts, the door to opportunity is not necessarily closed. The Dominican Republic is a case in point. Economic conditions in that country are less than positive. Yet, by working through the private sector, U.S. lumber and plywood producers have been very successful in exporting wood products to that country—successful to the point that the Dominican Republic is one of our most promising Latin American export markets. A growing tourist industry is projected to generate $330 million worth of construction in the country. This growth presents significant opportunities for U.S. wood products. Wood is already being considered for one 40-unit tourist villa.

High tariffs are another reason for not looking to Latin American markets. We can all take out our calculators and figure what the high tariffs do to our efforts to provide a low-cost housing solution. But when it comes to tariffs in developing countries, it's important to remember one point: nothing is set in concrete. (Of course, as producers of wood products, we ought to know that anyway!) Tariffs can be exonerated on certain products such as lumber and plywood if the governments of those countries declare housing a national priority. Tariffs can be changed, but it takes a lot of effort on our part to convince officials that wood products from the United States are the answer to housing shortages.

Then there's the matter of local traditions and preferences for other materials. One of the most common reasons we hear for not building wood houses in South America is that the people won't accept wood housing. They have lived in concrete or masonry houses since Bolivar and Pizarro, and concrete is what they want today. We disproved that theory in Peru. When we initiated work on fifteen demonstration units in Lima, the local officials warned us that the people wouldn't buy wood houses. But when the finished houses were placed on the market by the Peruvian Housing Authority, more than 2,490 families applied to purchase the houses. In fact, more families applied to buy the wood houses than the comparable brick ones.

Our demonstration houses have also drawn a positive reaction in other countries. There were those who thought the people of Jamaica would prefer concrete block shacks over wood frame construction. But after we built six low-income houses there, the Jamaicans expressed interest in twenty school buildings and more houses—projects that carry a U.S. export value of $500,000 to $700,000.

Cost is another negative argument frequently raised by the naysayers. Obviously, we have to be able to provide a cost-effective solution if we are to compete in foreign housing markets. Our product cost alone is usually competitive. Where we run into problems is in the shipping. Reduced shipping traffic to and from Latin American countries has escalated freight rates. Of course, we would probably ship lumber and plywood breakbulk. However, to simplify the shipping cost example, if you use a 40-foot container from Seattle to Valparaiso, Chile, it will cost $3,500, more than the material costs for one house. By comparison, container rates to the busier Asian ports of Hong Kong and Tokyo are under $1,000.

High tariffs and shipping costs notwithstanding, we have proved that we can still compete. Ten houses constructed of U.S. lumber and plywood were completed in Santiago, Chile in 1986. A local contractor familiar with U.S. wood systems was able to complete them, without U.S. supervisors, in record time by Chilean standards and at 24% less cost than comparable houses constructed with local materials and methods.

Figure 1. Demonstration house under construction in Peru.

Sometimes, however, lower costs may not open a market because of the need to convert local currency to dollars. It is often easy for the country to pay higher housing costs for local materials in local currency.

Local materials are often named as another justification for taking our business elsewhere. In most cases, concrete is the principal building material. But there are forests in Latin America, and many of the countries we've been working with do have their own plywood and lumber manufacturing facilities. These local manufacturers argue that their governments should use local products rather than importing U.S. materials. But despite arguments to the contrary, Latin American plywood and lumber manufacturers have not established themselves as the source for materials to overcome a five million housing unit deficit. They also don't manufacture products to consistent quality standards, nor do they hold to the treating standards we have in the United States, which would allow them to guarantee protection against termites and moisture.

FINANCING

A more serious obstacle is financing. We can overcome local traditions and predispositions against wood housing. And we can overcome misconceptions that local plywood and lumber manufacturers have the production capability to meet housing demands. We can even provide a quality, cost-competitive package. But without favorable financing, we have no means of getting our products to the market.

The most important means of financing wood product exports are GSM-102 and GSM-103, offered through the U.S. Department of Agriculture. The Export Credit Guarantee Program, commonly referred to as GSM-102, provides financing under a guarantee program that has terms of six months to three years. The finance terms are based on prevailing commercial bank rates.

To procure GSM-102 financing, either a U.S. exporter or the importing country submits a request form. When the request is received, the agricultural attaché resident in the buying country validates the request, based on the need for the commodity. The attaché's findings are forwarded to the USDA, where final approval on the guarantees may be given. Once approval is given, the exporter contracts with the buyer, registers the sale, and pays a guarantee fee to the USDA. The importer then must open an irrevocable letter of credit in favor of the exporter and drawn on an approved bank. The USDA may ask for an additional guarantee from the buyer's government that foreign exchange will be available to meet the obligation at maturity. The negotiated interest rates are usually commercial bank rates. Finally, the exporter assigns the USDA guarantee to the U.S. commercial bank and gets the proceeds from the sale once the terms of the letter of credit have been met. In the case of default, the commercial bank is paid by the Commodity Credit Corporation under the terms of the guarantee.

GSM-103 is similar to GSM-102 except that it extends the line of credit to ten years. This extra time is critical in the development of housing projects.

Both GSM-102 and GSM-103 are designed to cover agricultural commodities. Certain agencies are arguing that wood products should not be classified as agricultural commodities. The National Advisory Council, a committee composed of representatives from the various executive agencies, is objecting to GSM-102 and GSM-103 financing for wood products. The Treasury Department, the U.S. Trade Representative's Office, and State, Commerce, and other departments oppose the financing. They argue that plywood and other processed wood products are manufactured products rather than agricultural products and that export credit for manufactured products extending longer than the normal commercial terms is an export subsidy.

The U.S. wood products industry, the U.S. Department of Agriculture, and the U.S. Congress maintain that GSM-102 and GSM-103 are not subsidies and that wood products, including plywood, are agricultural commodities.

Figure 2. Completed demonstration house units in Peru.

It's all a matter of terms and definitions. But simple as it sounds, no single definition can have such far-reaching implications for U.S. wood product exporters. Because the resolution of GSM-102 and GSM-103 financing is so critical to our industry, we've formed a coalition designed specifically to make it clear that wood products are rightfully classified as agricultural commodities: U.S. Coalition for Forest Product Exports. The coalition is composed of the American Plywood Association, the Hardwood Export Trade Council, the National Association of State Foresters, the National Forest Products Association, the Southern Forest Products Association, and the Western Wood Products Association.

We all have to be 100% dedicated to the favorable resolution of wood export financing—not only for our efforts in Latin America but also for Asia and the Middle East. We have already shown that we can overcome other barriers to trade with developing countries. But until we establish finance programs, our other efforts and achievements are fruitless.

THE FUTURE

Establishing trade with Latin America is a long, hard process. For all the reasons I listed above, opening doors to trade requires time, patience, and the ability to see beyond today's roadblocks.

Argentina is a good example of the time investment required to generate significant orders for U.S. wood products. We started talking with Argentine officials at the 1984 Inter-American Housing Conference, held in Seattle. As a result of contacts made at the conference, U.S. Congressman Don Bonker led a trade mission to Argentina in 1986. We met with Housing Secretary Juan Maisterrena, who, encouraged by the trade mission, returned the visit with a trip to the Pacific Northwest in January 1987.

Now we are moving ahead on Secretary Maisterrena's proposal to build 600 wood houses in Buenos Aires by the end of 1987. Argentina is also proposing construction of 1,400 units in Buenos Aires by 1990; 1,000 units in Tierra del Fuego; and as many as 300,000 units in Viedma, the new capital city.

Yes, this is all taking a lot of time. But we are working toward tangible results that make the effort worthwhile. To realize those results, we must continue working for slow but sure progress.

Recently, while responding to Brazil's unilateral suspension of interest payments, Federal Reserve Board Chairman Paul Volcker said U.S. banks are much less vulnerable to international debt programs than they were four years ago, because they have substantially increased their capital. He said that there has been more progress in Latin America toward fundamental economic reform than at any time in the last fifty years. He voiced concern that the United States is showing signs of battle fatigue—that we are giving up.

With the opportunity to help answer a critical need for five million housing units a year, we can't afford to give up on Latin America. The need for housing is real, and we as an industry have the products and the know-how to answer that need.

Chapter 4

An Ocean Freight Cost Analysis for Chilean Forest Product Exports

ARNALDO JELVEZ CAAMAÑO

Chile is becoming an important exporter of softwood products. Its forestry economics is based mainly on fast-growing plantations of radiata pine, which account for over 90% of the total production. Chilean forest product exports amounted to a value of U.S.$382.7 million in 1984, or 10.3% of the country's total exports. The percentage decreased to 8.8% in 1985 but rose again to 10% in 1986.

Based on 1984 figures, Chile sold wood products to sixty countries. Exports to Asia, South America, and Europe accounted for 90% of total sales in value terms (INFOR 1985).

The objective of this study is to estimate the importance of ocean freight cost in "delivered prices" (defined as free on board price plus ocean freight costs) for Chilean forest products. Because of Chile's remoteness from the major world forest product consumers, ocean freight costs are a crucial and sometimes decisive variable in competing in the international market on a c.i.f. (cost, insurance, and freight) price basis. This variable is having more influence considering the decreasing f.o.b. (free on board) forest product prices in recent years.

The representative forest products used for this study were bleached pulp, radiata pine sawnwood, newsprint, and logs of radiata pine, which together accounted for 61% of the total forest value exports in 1984, 64% in 1985, and 63% in 1986.

For each product the relevant markets were identified. Results are given for the most important one in terms of volume as well as for the "world market." Simple averages are provided of the relative share of the ocean freight cost to all countries to which the products were exported.

The data were obtained from the Instituto Forestal (INFOR) of Chile and correspond to the period 1981-85. Values are given in real terms. Ocean freight cost shares average 11.8% (range 9 to 21%) for all the markets in bleached pulp. Newsprint shows an average of 11.6% (range 8 to 28%). In sawnwood the average is 31.4% (range 8 to 39%). Logs of radiata pine show the highest shares with an average of 42.2% (range 29 to 51%). Supply and demand determine the ocean freight cost at the international level. However, cartellike agreements have oligopolistic impacts. This means that the Chilean markets are not only affected by economic efficiency but also by external and uncontrollable factors (Centro de Estudios Navieros 1985).

The main incentive for carrying out this study was the scarcity of information on the topic. The only reference available (Hartwig 1982) mentions that for native species the ocean freight cost in general should not exceed 30% of f.o.b. values. The present study shows that this percentage for sawnwood and export logs is exceeded. For this reason, it would seem that, together with efforts to decrease production costs, attempts have to be made to negotiate better rates in freight costs. The impact of these factors will be more crucial in the near future because the supply of Chilean forest products to tne world will increase substantially over the next ten years (INFOR 1985, Flora 1986).

METHODOLOGY

The source of information used to prepare this paper was the monthly bulletin of exports published by the Instituto Forestal, Santiago, Chile. Each report segregates for product and country of destination the exporters, volume, f.o.b. price, and ocean freight cost. From these figures a simple monthly average for product and market was derived. Considering the amount of information to be processed, and given that the individual volumes were not significantly different, a simple average was used. Monthly figures allowed obtaining the annual values. Average values for each market were obtained using the classification of the Instituto Forestal. The different markets are grouped geographically. Nominal values were extracted using the wholesale price index of the United States, with 1980 as base 100.

BRIEF OVERVIEW OF THE PORT COMPLEX IN CHILE

The port complex of Region VIII region in Chile includes the ports of San Vicente, Lirquén, and Talcahuano. These ports account for approximately 10% of the nation's cargo and for 95% of the total forest exports.

The ports of San Vicente and Talcahuano are administered by the government through EMPORCHI (Empresa Portuaria de Chile), while the port of Lirquén is privately managed. Even so, rates applied to the different stages of the port movements for forest products are similar.

Ships of lines that do not belong to a cartel agreement are generally cheaper; however, their role in exporting Chilean forest products is small.

Ocean freight cost is affected mainly by fuel costs, type of ship, labor costs, specific legal considerations of the importing nation, cargo reserves, or other flag restrictions. However, distance is another important variable (Wisdom and Jones 1986). As a reference, distances in terms of number of shipping days are given in Table 1 for the main Chilean markets for average products and ship types. Bulk carriers are generally used to ship lumber and logs, and liner freighters are used for bleached pulp and newsprint.

Table 1. Average shipping time from Chile to principal markets.

Market	Shipping Time (days)
South America	8
Central America	18
North America	22
Europe	25
Middle East	25
Japan	30
Taiwan	35

A recent study in Chile shows that the capacity of the ports is sufficient to cover actual demand now and for the next four years. (Secretaría Regional de Planificación y Coordinación, Región del Bíobio 1985). However, after this period new investments will be required, and some are already under way.

RESULTS

Bleached Pulp

In 1984, bleached pulp generated a total of $103.3 million f.o.b. Europe with a 55% share, South America with 32%, and Asia with 12% are the largest markets. In 1985 there was an increase of 18% of the total sales to Asia and a decrease of 21% to South America.

Eighty-nine transactions were sampled. In the world market the ocean freight cost has an average share of 11.8% with a range between 9.8 to 14% (see Table 2). The South American market shows the lowest percentages with an average of 10.4%, and Asia the highest with 15.5%.

Table 2. Percentage share of ocean freight cost in real values, bleached pulp.

Market	1981	1982	1983	1984	1985
World market	9.8	11.8	14.0	11.5	12.0
Europe	10.2	10.4	12.0	10.3	13.3
South America	9.2	10.9	11.5	9.8	10.5
Asia	10.4	16.7	20.8	16.0	13.8
North America	--	--	12.6	12.0	13.1
Central America	--	--	11.9	--	--

Figures for freight costs show an interesting stability for world and European markets (see Figures 1 and 2). In the world market this value is around U.S.$51/m^3. In 1985 a reduction of 9% with respect to the average figures was observed. Average freight cost to the European market is U.S.$45/m^3, showing a downward trend compared with 1981 values.

In 1983 in the world market the highest value for ocean freight cost was reached, while in the same year the f.o.b. price for pulp was lowest (U.S.$323.4/m^3) (see Figure 1).

Considering the stability in the ocean freight costs for the period, the changes in the percentage shares are directly caused by the decreasing f.o.b. prices for bleached pulp since 1981.

Newsprint

In 1984, newsprint exports amounted to U.S.$39.8 million f.o.b. South America was the most important market, with 72% of the sales. Seventy transactions were sampled.

The share of the freight cost reached an average of 16.6% (range is 13 to 19%)(see Table 3). The world market showed many fluctuations in the relative share of the ocean freight cost (see Figure 3). The highest value was reached in 1983-84 when the highest freight cost and the lowest f.o.b.

Table 3. Percentage share of ocean freight cost in real values, newsprint.

Market	1981	1982	1983	1984	1985
World market	13.1	15.2	18.9	17.8	18.1
South America	13.1	13.8	15.3	15.7	13.9
Asia	--	25.7	28.3	27.3	25.4
Central America	--	7.9	12.6	13.3	12.4
Europe	--	15.9	19.8	13.2	--
North America	--	18.3	18.3	16.8	21.2
Africa	--	--	--	4.9	17.9
Oceania	--	--	20.5	--	--

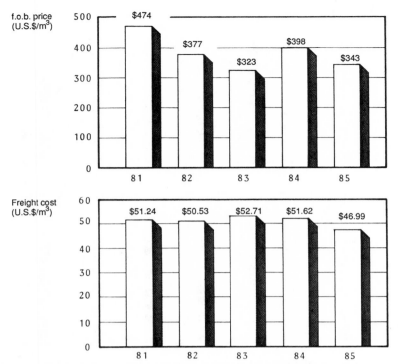

Figure 1. Bleached pulp (radiata pine) value and freight cost in the world market, 1981-85.

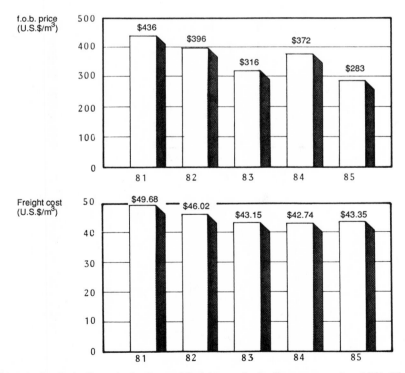

Figure 2. Bleached pulp (radiata pine) value and freight cost in the European market, 1981-85.

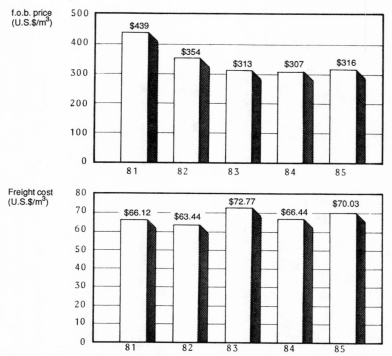

Figure 3. Newsprint (radiata pine) value and freight cost in the world market, 1981-85.

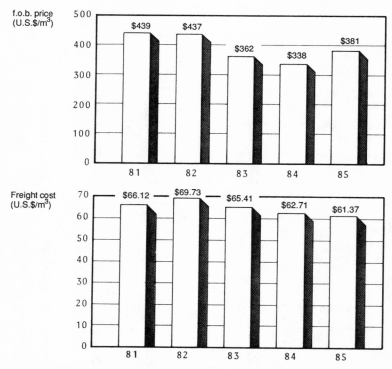

Figure 4. Newsprint (radiata pine) value and freight cost in the South American market, 1981-85.

price occurred almost simultaneously. The South American market showed smaller fluctuations (Figure 4), with values ranging between 13 to 15.7%.

The second relevant market for newsprint is Asia, with a range between 25 and 28%, which almost doubled the average values for the world market (see Table 3). This is explained mainly by the increasing freight costs with respect to the f.o.b. prices. In this market, ocean freight costs have values 34% over the average world market.

Sawnwood

Exports of sawnwood generated a total of U.S.$61 million in 1984. Asia and South America, sharing 40% and 31% respectively, are the most important markets for Chilean sawnwood. In the world market the freight cost for sawnwood reaches 31.4% as an average (with a range of 30 to 34%) (see Table 4). The highest value of 34% was reached in 1981. Freight costs and f.o.b. prices show a continous decrease in the world market, with freight costs more affected (see Figure 5).

Table 4. Percentage share of ocean freight costs in real values, radiata pine sawnwood.

Market	1981	1982	1983	1984	1985
World market	34.0	30.0	31.8	31.3	30.3
Asia	39.4	31.6	34.7	31.2	37.6
South America	32.9	30.1	27.2	24.8	23.7
Africa	32.7	27.6	28.2	33.4	32.3
North America	32.6	24.8	24.3	20.3	28.1
Central America	26.2	24.9	22.3	25.9	22.5
Europe	38.0	39.7	42.4	44.1	--

In the Asian market, a decrease in freight costs can be observed but not in f.o.b. prices, which show significant fluctuations (Figure 6). In this market the highest percentage was reached in 1981 with 39.4% and the lowest in 1984 with 31.2%.

The second important market for Chilean sawnwood is South America. In this market the share of the freight cost has decreased since 1981. For instance, the share in 1985 was 42% of the average value paid in 1981. On the other hand, only in 1984 did the f.o.b. price per cubic meter decrease in relation to the world market. The other figures indicate increasing values in real terms. These two effects explain that the ocean freight cost share went from 32.9% in 1981 to 23.7% in 1985 (see Table 4).

Export Logs

Export of logs generated U.S.$29.4 million in 1984. This figure increased to $39.4 million in 1985. Over 98% of the logs exported go to the Asian market. Results obtained for the world and Asian markets were very similar (Table 5), because of the low importance of other markets as far as this product is concerned. The average share of freight costs is 42.4% of the total costs and the range is between 36 and 46%. The decreasing trend in f.o.b. prices for the period is clear (see Figures 7 and 8). For this product the same trend is observed for freight costs.

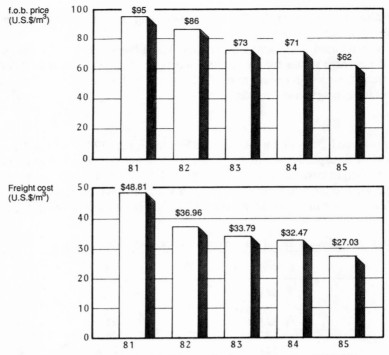

Figure 5. Sawnwood (radiata pine) value and freight cost in the world market, 1981-85.

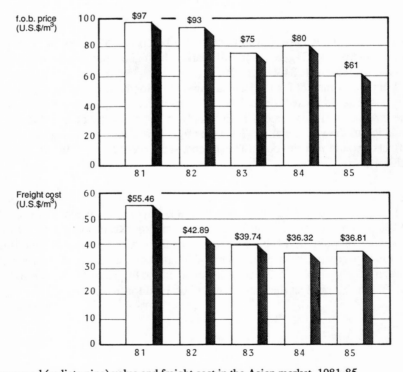

Figure 6. Sawnwood (radiata pine) value and freight cost in the Asian market, 1981-85.

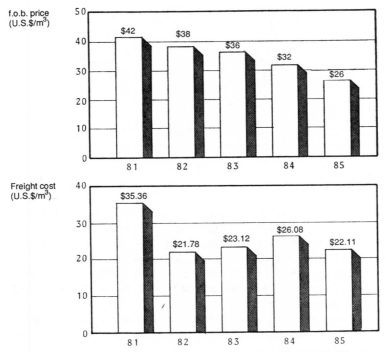

Figure 7. Export log (radiata pine) value and freight cost in the world market, 1981-85.

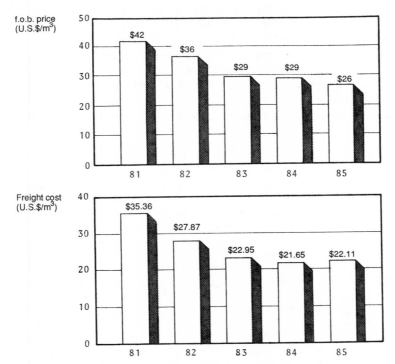

Figure 8. Export log (radiata pine) value and freight cost in the Asian market, 1981-85.

Table 5. Percentage share of ocean freight cost in real values, export logs.

Market	1981	1982	1983	1984	1985
World market	45.9	36.3	39.0	45.2	45.9
Asia	45.9	43.4	44.1	43.1	45.9
Africa	--	4.8	--	50.0	--
South America	--	28.9	29.2	39.7	--
Europe	--	--	--	50.7	--
Central America	--	38.2	--	--	--

CONCLUSIONS

The ocean freight cost represents an important proportion of the "delivered price" for the Chilean forest product exports. The average percentages are 12% for bleached pulp, 17% for newsprint, 31% for radiata pine sawnwood, and 42% for export logs. The outlook for pulp is one of stability in real ocean freight costs, which averaged about U.S.$51/m^3 during 1981 to 1985. In newsprint, the ocean freight cost share fluctuated from 13% in 1981 to 18% in 1985. In radiata pine sawnwood, a sustained reduction in freight costs took place from U.S.$49/m^3 in 1981 to U.S.$27 in 1985. The same trend was observed for f.o.b. prices.

Since the lowest f.o.b. prices are recorded for pulp and newsprint, these are the most affected by the ocean freight costs; for example in newsprint, increases of over 34% in the average values in the world market are observed.

For the forest products considered in this study, except for newsprint, a clear decreasing tendency in freight cost is observed. Unfortunately, the same does not hold for the f.o.b. prices.

If we estimate the total value of exports for the four products in this study using average prices, we come to $233.5 million in 1984 nominal values. Estimating in the same manner the value paid in ocean freight cost, a total of $77 million is reached. This means that for each dollar paid (referred to f.o.b. and c.i.f. terms), approximately 25 to 33% is paid in ocean freight. Considering that f.o.b. price includes not only the production cost but also internal transportation, product storage, stowage, and others, for some products and markets the ocean freight cost paid is greater than the internal production cost of the product.

Even so, the future of the international market for Chilean forest products looks promising.

REFERENCES

Banco Central de Chile. 1984. Indicador de comercio exterior, Santiago, Chile. Bol. Diciembre.

Banco O'Higgins y Pro-Chile. 1985. Seminario regional para el desarrollo de las exportaciones. Apuntes, Concepción.

Centro de Estudios Navieros. 1985. Los usuarios y las conferencias de fletes. Segundas Jornadas de estudio sobre Transporte Marítimo, Santiago, Chile.

Comisión Económica para América Latina. 1984. Transporte marítimo en la Región Austral de Chile. Documento distribución restringida.

CONAF. 1984. Boletines mensuales de Mercado.

Flora, D. F. 1986. An equilibrium model of Pacific Rim trade in small softwood logs. Can. J. For. Res. 16:1000-1006.

Hartwig, F. 1982. Guía para la exportación de madera de especies del bosque nativo chileno a países europeos. Documentod de Trabajo 45.

INFOR. 1985. Exportaciones Forestales Chilenas. Serie Informática 26. Santiago, Chile.

Secretaría Regional de Planificación y Coordinación, Región del Bíobio. 1985. Estudio de demanda de infraestructura de transporte en la VIII Región. Informe final.

Wisdom, H. W., and K. S. Jones. 1986. The determinants of ocean freight cost for forest products. Can. J. For. Res. 16:701-709.

Chapter 5

European, Soviet, and North American Trade Patterns in Forest Products with Middle Eastern and North African Countries

T. J. PECK

After centuries of relative quiescence, the countries of the Middle East and North Africa found their way to the center of the world's geopolitical stage in the early 1970s, when the Organization of Petroleum Exporting Countries (OPEC) sharply raised the price of crude oil on the world market. The reverberations of this move are still being felt. And while it could be said with hindsight that the world economy was almost inevitably heading sooner or later for a major structural shift, after a period of remarkable prosperity and growth, it was the oil price shock that triggered the world economic recession of 1974-75, which has been followed by a decade of uncertainty, loss of confidence, and appreciably slower growth in most parts of the world. Another major effect was a redistribution of funds from the oil-importing to the oil-exporting countries. Some of these funds have been employed for accelerating the social and economic development of the oil-rich countries, including major construction programs: housing, hospitals, schools, industries, roads, harbors, airports, and so on.

To meet the needs of these programs, the countries concerned have had to import large quantities of goods and services. Forest products have figured prominently among the former: sawnwood and wood-based panels for the construction programs; paper to meet the strongly increasing demand arising from higher standards of living, improved education, and greater literacy. The countries of the Middle East and North Africa are today among the very poorest, so far as their forest resources are concerned. Nearly all their needs for forest products, apart from fuelwood, have therefore had to be met by imports.

This paper considers the trends since the early 1970s in the exports of selected forest products from Europe, the U.S.S.R., and North America to nineteen countries (North Africa: Algeria, Egypt, Libya, Morocco, Tunisia; Middle East: Bahrain, Iran, Iraq, Israel, Jordan, Kuwait, Lebanon, Oman, Qatar, Saudi Arabia, Syria, United Arab Emirates, Yemen A. R., and Democratic Republic of Yemen). Many of these countries are oil exporters and, as a result of the oil price jumps of 1973-74 and 1979-80, found themselves with huge funds available for economic and social development, among other things. Their import needs rose rapidly, not least for forest products, since domestic sources (forests and industries) of these products are small or nonexistent. Sawn softwood, paper and paperboard, plywood, and sawn hardwood, in that order in terms of value, are the main forest product imports of these countries. The total value of their forest product imports in 1984 was over U.S.\$3 billion, or about 5.5% of world trade. The sharp drop in oil prices in 1985-86 had an immediate effect on funds available for purchasing imports, as shown by exporting countries' results for 1986, now (February 1987) becoming available. It may be concluded that until the oil-producing countries of the Middle East and North Africa are able to diversify their economies—a difficult and long-term enterprise—so as to be less reliant on oil revenues, their import demand for forest products will continue to be linked rather closely to developments in the world oil market. This makes it virtually impossible to find a reasonable set of assumptions on which to base forecasts of future consumption and import trends.

DEMOGRAPHIC AND ECONOMIC BACKGROUND

The region considered here ranges from the Indian Ocean to the east, the borders of Afghanistan, the U.S.S.R., Turkey, and the Mediterranean to the north, the Atlantic Ocean to the west, and the Sahara to the south (Figure 1). The selection of the countries covered by the paper was arbitrary. The majority are Arab speaking; several are members of OPEC; all are poor in forest resources; the climate in most of the area is dry and hot. Basic information on their land area, populations, and gross national products is given in Table 1.

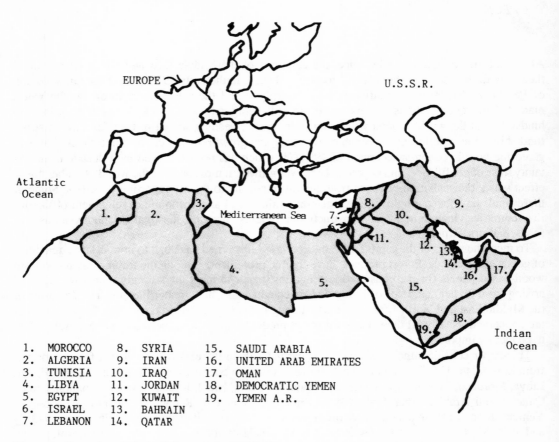

1.	MOROCCO	8.	SYRIA	15.	SAUDI ARABIA
2.	ALGERIA	9.	IRAN	16.	UNITED ARAB EMIRATES
3.	TUNISIA	10.	IRAQ	17.	OMAN
4.	LIBYA	11.	JORDAN	18.	DEMOCRATIC YEMEN
5.	EGYPT	12.	KUWAIT	19.	YEMEN A.R.
6.	ISRAEL	13.	BAHRAIN		
7.	LEBANON	14.	QATAR		

Figure 1. Countries of North Africa and the Middle East.

In the early 1980s, the nineteen countries had a combined population of around 200 million, 15% less than that of the United States, which was 235 million. The total land area amounts to over 11 million square kilometers, which is about 20% more than that of the United States. Total gross national product (GNP) of the nineteen countries was estimated at U.S.$581 billion or about $2,925 per capita. The corresponding figures for the United States were $3,276 billion and $14,110 per capita. It is important to note, however, that the countries in the region range from among the very

Table 1. Land area, population, and gross national product in North African and Middle Eastern countries in 1983.

Countries	Land Area (1,000 km^2)	Population (millions)	Gross National Product Total (million U.S.$)	Gross National Product Per capita (U.S.$/ capita)
Algeria	2,382	20.57	51,840	2,520
Egypt	1,001	45.20	27,080	700
Libya	1,760	3.40	28,830	8,480
Morocco	447	20.80	15,810	760
Tunisia	164	6.90	8,900	1,290
North Africa (5 countries)	5,754	96.87	132,460	1,370
Bahrain	1	0.40	4,200	10,510
Iran	1,648	42.51	161,540	3,800
Iraq	435	14.66	41,780	2,850
Israel	21	4.18	20,660	5,370
Jordan	98	3.24	5,310	1,640
Kuwait	18	1.70	21,330	17,880
Lebanon	10	2.62	2,490	950
Oman	300	1.18	7,460	6,250
Qatar	11	0.28	5,940	21,210
Saudi Arabia	2,150	10.44	127,680	12,230
Syria	185	9.60	16,900	1,760
United Arab Emirates	84	1.21	27,670	22,870
Yemen A.R.	195	7.60	4,180	550
Yemen, Democratic Republic	333	2.01	1,000	520
Middle East (14 countries)	5,489	101.63	448,140	4,410
Total (19 countries)	11,243	198.50	580,600	2,925

Sources: FAO *Yearbook of Forest Products* 1984; UNDP *Annual Report* 1985.

poorest in the world, with a GNP of little more than $500 per capita, to the richest. Three countries had a GNP per capita higher than that of the United States—Kuwait, Qatar, and the United Arab Emirates—the last with a figure of $22,870 per capita for its population of 1.2 million.

The two largest economies are Iran and Saudi Arabia, followed by Algeria, Iraq, Libya, United Arab Emirates, and Egypt. Even so, only the first two are larger in terms of total GNP than Switzerland.

Very roughly, land area and population are divided equally between the five North African countries and twelve Middle Eastern ones; but the latter account for over three-quarters of the total GNP of the area.

FOREST RESOURCES AND DOMESTIC WOOD SUPPLY

It has already been mentioned that the Middle East and North Africa are poor in forests. Table 2, based on information collected by FAO, puts the situation in figures. For the sixteen countries listed, the total area of forest and other wooded land is estimated at not quite 30 million hectares. This represents only 2.7% of the total land area and amounts to 0.14 ha per capita. Equivalent figures for the United States are 298 million ha, 32.7%, and 1.27 ha per capita, respectively.

Table 2. Area of forest and other wooded land in North Africa and Middle Eastern countries.

Countries	Total Forest and Other Wooded Land			Closed Forest		Other Wooded Land	
	Area (1,000 ha)	Share of total land area (%)	Area per capita (ha)	Area	of which: Coniferous (1,000 ha)	Area	of which: Coniferous (1,000 ha)
Algeria	2,990	1.3	0.14	490	410	2,500	2,100
Egypt	31	--	--	1	--	30	--
Libya	480	0.3	0.13	70	50	410	280
Morocco	5,200	11.6	0.24	400	200	4,800	1,100
Tunisia	750	4.6	0.11	400	130	350	100
North Africa (5 countries)	9,451	1.6	0.09	1,361	790	8,090	3,580
Iran	12,400	7.5	0.35	1,900	20	10,500	2,500
Iraq	1,910	4.4	0.12	30	20	1,880	80
Israel	100	4.8	0.02	80	55	20	10
Jordan	82	0.8	0.02	54	18	28	10
Kuwait	15	0.8	0.01	--	--	15	15
Lebanon	85	8.2	0.03	35	10	50	15
Saudi Arabia	1,200	0.6	0.10	--	--	1,200	600
Syria	420	2.3	0.04	90	40	330	10
United Arab Emirates	3	--	--	3	--	--	--
Yemen A.R.	2,500	12.8	0.37	--	--	2,500	--
Yemen, Dem. Rep.	1,600	4.8	0.75	--	--	1,600	--
Middle East (11 countries)	20,315	3.9	0.19	2,192	163	18,123	3,240
Total (16 countries)	29,766	2.7	0.14	3,553	953	26,213	6,820

Source: FAO *World Forest Resources* 1980.

The highest forest coverages are reported from the Yemen Arab Republic (12.8%) and Morocco (11.6%), but in these countries much of what comes under the global term "forest and other wooded land" is various forms of scrub. In fact, for the sixteen countries, over 26 million ha, or 88% of the total, comes under the category of "other wooded land" (i.e., open woodland with less than 20% tree crown cover, scrub, brushland, etc.). Only 3.55 million ha are classified as closed forest, of which 1.9 million ha are in Iran and a further 1.3 million ha are divided between Algeria, Morocco, and Tunisia.

Although there are exceptions, the quality of national statistics on the forest resource among the nineteen countries leaves much to be desired. This applies to other components of the forest and forest products sector, including domestic wood supply (removals) and trade. A number of reasons may be advanced for this situation, but essentially it comes down to the fact that forests and wood are not major or essential elements of the national economies; furthermore, wood is not a familiar product, except in its traditional use as fuel. Therefore, not too much reliance should be placed on either the volumes reported by the countries concerned or on the assortment composition, either of domestic supply or of imports. This is not intended as a criticism: in government as in business, everything has a priority. In most of the countries under review, the statistical coverage of the forest and forest products sector is generally not considered to be of particular importance.

With this in mind, the figures of roundwood production (removals) in Table 3 may be noted. Reported removals in twelve countries in 1983, the latest year for which reasonably complete data or estimates are available, amounted to 16 million m^3. Leading producers were Iran, Tunisia, Egypt, Algeria, and Morocco. Of the total, over 11 million m^3, or 65%, were fuelwood removals, and the remainder various assortments of industrial wood. The proportion of industrial wood in the total is strongly influenced by the figures for Iran, which are in fact estimates; and, given the estimated volumes of output of processed wood products in Table 4, it seems very unlikely that Iran had removals of industrial wood of over 4 million m^3 in the early 1980s. Excluding Iran, the share of industrial wood in the remaining eleven countries in Table 3 was 13%, and of fuelwood 87%, a breakdown not inconsistent with the quality of the forest resource and the level of development in the forest industry sector.

It can be said, therefore, that with the doubtful exception of Iran, domestic supplies are contributing in only a limited way to the requirements for industrial wood and its products in the countries of the Middle East and North Africa.

Table 3. Reported roundwood removals in 1983 in North Africa and Middle Eastern countries.

Countries	Total Volume (1,000 m^3)	Total Change on 1970 (%)	Industrial Wood Volume (1,000 m^3)	Industrial Wood Share of total removals (%)	of which: Coniferous Volume (1,000 m^3)	of which: Coniferous Share of total industrial wood (%)
Algeria*	1,680	+36	219	13	148	68
Egypt*	1,910	+34	89	5	--	0
Libya*	631	+45	95	15	--	0
Morocco*	1,685	+17	565	34	99	18
Tunisia	2,692	+25	97	4	27	28
North Africa (5 countries)	8,598	+29	1,065	12	274	26
Iran*	6,727	−7	4,376	65	--	--
Iraq*	131	+42	50	38	--	--
Israel*	118	+37	107	91	62	58
Jordan*	9	+80	4	44	--	--
Lebanon	469	+22	25	5	17	68
Syria	44	−39	34	77	8	24
Yemen, Dem. Rep.*	270	+36	--	0	--	0
Middle East (7 countries)	7,768	−4	4,596	59	--	--
Total (12 countries)	16,366	+11	5,661	35	--	--

Source: FAO *Yearbook of Forest Products* 1984.
*Unofficial figures or estimates.

Table 4. Reported production of forest products in 1983 in North Africa and Middle Eastern countries.

Countries	Sawnwood		Wood-based Panels		Wood-pulp	Paper and Paperboard		
	Total	of which: Coniferous	Total	of which: Plywood		Total	News-print	Printing & writing
	(1,000 m^3)					(1,000 m.t.)		
Algeria	13*	8*	50*	23*	--	77*	--	30*
Egypt	--	--	37	8	--	110*	--	39*
Libya	31*	--	--	--	--	5*	--	--
Morocco	149	94*	105*	50*	80*	82*	--	17*
Tunisia	3	2*	99	24*	--	23*	--	23
North Africa (5 countries)	196	104	291	105	80	297	--	109
Iran	163*	--	171	16*	--	78*	--	45*
Iraq	8*	--	2*	--	--	28*	--	9*
Israel	--	--	138	71*	--	157	2*	67
Jordan	--	--	--	--	--	5	--	--
Lebanon	33	22*	46*	34*	--	45*	--	--
Syria	9	7*	27*	8*	--	3*	--	--
Middle East (6 countries)	213	29	384	129	--	316	2	121
Total (11 countries)	409	133	675	234	80	613	2	230

Source: FAO *Yearbook of Forest Products* 1984.
*Unofficial figures or estimates.

IMPORTS OF FOREST PRODUCTS

All of the countries under review, therefore, rely to a considerable extent—and in some of them exclusively—on imported forest products. Tables 5 (value) and 6 (volume) show the basic import data for 1984, taken from the FAO *Yearbook of Forest Products*. The earlier warning about the quality of statistics needs to be repeated for the case of imports. Bearing this reservation in mind, it may be estimated that the total value of imports in 1984 by the seventeen countries shown in Table 5 was U.S.$3.01 billion, or 5.4% of the world total. Of the total value of imports of the seventeen countries, sawnwood accounted for 42.3%, paper and paperboard for 28.7%, wood-based panels for 17.9%, roundwood for 8.2%, and woodpulp for only 3.0%. Compared with the world distribution of imports by product group, the countries of North Africa and the Middle East have a substantially greater part of their imports in the form of sawnwood and wood-based panels and a much lower proportion in woodpulp and paper and paperboard. Thus, compared with their 5.4% share of world forest product imports, those of the product groups were as follows in 1984: roundwood, 3.1%; sawnwood, 11.1%; wood-based panels, 11.6%; woodpulp, 1.0%; and paper and paperboard, 3.9%.

In terms of importance as purchasers of forest products from overseas, the leading countries in 1984 were: Egypt, U.S.$664 million; Saudi Arabia, $512 million; Algeria, $378 million; Iran, $263 million; Israel, $187 million; Iraq, $158 million; and Morocco, $127 million.

Table 6 shows the volumes of forest product imports by product and country in 1984. The out-

Table 5. Value of imports of forest products by North African and Middle Eastern countries in 1984.

Countries	Value of Imports (U.S.$ million)						% of Total				
	Total	Roundwood	Sawnwood	Wood-based panels	Woodpulp	Paper and paperboard	Roundwood	Sawnwood	Wood-based panels	Woodpulp	Paper and paperboard
Algeria	378.4	35.1	188.4	47.6	29.8	77.4	9.3	49.8	12.6	7.9	20.4
Egypt	663.6	58.1	346.7	77.6	--	181.3	8.8	52.2	11.7	--	27.3
Libya	110.8	10.0	56.8	26.2	--	17.8	9.0	51.3	23.6	--	16.1
Morocco	126.6	22.4	51.7	4.4	9.2	38.9	17.7	40.8	3.5	7.3	30.7
Tunisia	115.8	3.5	54.9	11.2	7.6	38.6	3.0	47.4	9.7	6.6	33.3
North Africa (5 countries)	1,395.2	129.1	698.5	167.0	46.6	353.9	9.3	50.1	12.0	3.3	25.4
Bahrain	46.4	14.5	10.2	15.0	--	6.7	31.3	22.0	32.3	--	14.4
Iran	263.3	21.7	62.2	35.1	20.4	123.8	8.3	23.6	13.3	7.8	47.0
Iraq	158.1	0.2	97.4	25.9	8.6	26.1	0.1	61.6	16.4	5.4	16.5
Israel	186.6	21.7	49.7	7.5	14.1	93.8	11.6	26.6	4.0	7.5	50.3
Jordan	77.8	5.3	18.7	12.2	--	41.6	6.8	24.0	15.7	--	53.5
Kuwait	114.1	1.5	32.3	45.4	--	34.9	1.3	28.3	39.8	--	30.6
Lebanon	65.7	1.2	24.6	3.5	--	36.4	1.8	37.5	5.3	--	55.4
Oman	53.5	2.4	25.2	19.9	--	6.0	4.5	47.1	37.2	--	11.2
Qatar	17.6	1.6	--	6.5	--	9.6	9.0	--	36.7	--	54.3
Saudi Arabia	512.3	40.3	192.0	188.7	--	91.2	7.9	37.5	36.8	--	17.8
Syria	111.0	4.4	56.4	11.6	--	38.6	4.0	50.8	10.4	--	34.8
Yemen, Dem. Rep.	11.3	2.5	5.7	1.5	--	1.5	22.6	50.3	13.7	--	13.4
Middle East (12 countries)	1,617.7	117.3	574.4	372.9	43.1	510.0	7.2	35.5	23.1	2.7	31.5
Total (17 countries)	3,012.9	246.4	1,273.0	539.9	89.7	863.9	8.2	42.3	17.9	3.0	28.7
WORLD	55,772.0	7,994.0	11,504.0	4,674.0	9,336.0	22,264.0	14.3	20.6	8.4	16.7	39.9

Source: FAO *Yearbook of Forest Products* 1985.
Note: Detail may not add to total, because of rounding.

Middle East and North Africa 59

Table 6. Volume of imports of forest products by North African and Middle Eastern countries in 1984.

Countries	Round-wood	Sawnwood Total	of which: Con-iferous	of which: Noncon-iferous	Wood-based Panels Total	of which: Plywood	Particle-board	Fiber-board	Wood-pulp	Paper and Paperboard Total	of which: News-print	Printing & writing	Other paper & board
				(1,000 m³)						(1,000 m.t.)			
Algeria	285	1,114	1,100*	6*	93	88*	3*	--	64	97	11	30*	56*
Egypt	302	2,078	1,857	213	182	176	--	--	--	385	57	113	215
Libya	74	219	219*	--*	75	72*	--	--	--	15*	6*	3*	6*
Morocco	144	423	376	45	14	4	--	--	23	76	7	2	67
Tunisia	25	315	280	34	20	--	--	5*	18	63	8	7*	48*
North Africa (5 countries)	830	4,149	3,832	298	384	340	--	--	105	636	89	155	392
Bahrain	79	45	--	45*	40*	43*	--	--	--	4*	--	2*	2*
Iran	118	305	300*	4*	88*	50*	35*	--*	41	200	23*	27*	150*
Iraq	6	407	336*	659	46	39*	4*	--	15	39*	10*	2*	27*
Israel	149	257	249	9	22	1	6*	10	27	173	45	16	112
Jordan	3	93	91*	2*	39	39	--	--	--	54	10	13	31
Kuwait	40	188*	188*	--*	118	104	11*	--	--	51*	15*	17*	19*
Lebanon	31	70	70*	--*	22	7*	14*	--*	--	64	5*	34 *	25*
Oman	8	192	1*	191*	66	70*	--	--	--	6*	--	4*	2*
Qatar	36	--	--	--	18*	18*	--*	--*	--	6*	--	3*	3*
Saudi Arabia	238	1,365	1,241	123	623	527	--	--	--	142	14	79	49
Syria	82	286	230	55	36	9	14	--	--	67	7	17	43
Yemen, Dem. Rep.	5	29*	8*	21*	2	2*	--	--	--	2*	1*	1*	--*
Middle East (12 countries)	795	3,237	2,714	519	1,120	909	84	--	83	808	130	215	463
Total (17 countries)	1,625	7,386	6,546	817	1,504	1,242	--	--	188	1,444	219	370	855

Source: FAO Yearbook of Forest Products 1985.
*Unofficial figures or estimates.

standing importance of sawn softwood among individual products is apparent, with plywood and sawn hardwood also imported in large volumes, as well as all categories of paper and paperboard.

The large number of asterisks (*) in Table 6 indicates the extent to which total import volumes are either unofficial figures or have had to be estimated on the basis of data for earlier years or by some other means. One way of cross-checking the general level of acceptability of the import figures is to compare them with information available from the forest product exporting countries of their trade with the countries of North Africa and the Middle East. In many cases, because of the importance of forest products for the economies and export earnings of these countries, and because of their familiarity with them, the quality of the export statistics is rather high.

Based on the common data collection system operated by the FAO Forestry Department (Rome) and the FAO/ECE Agriculture and Timber Division (Geneva), the latter is building up a TIMTRADE data base of trade flows between a majority of the countries in the world. So far only some years have been entered (1970, 1975, 1980, 1984) and the sixteen most important forest products. The decision was taken at the start of the work to base TIMTRADE on exporting countries' statistics as far as possible, for the reasons mentioned above.

Gaps were filled in, wherever necessary and possible, by using importers' figures. In this way, the trade flow coverage amounted for most products and years to well over 90% of the world total. It can be said, therefore, that while TIMTRADE is not fully comprehensive and, as with efforts by other organizations of a similar kind, has its weaknesses and drawbacks, it is proving to be a workable tool for trade analysis purposes.

For the present paper, use has been made of TIMTRADE to examine the volume and trends of trade in selected forest products between the countries of the ECE region (Europe, the U.S.S.R., North America) and thirteen countries of North Africa and the Middle East that are identified as importers. Some basic trade figures for 1984 from TIMTRADE are shown in Table 7 alongside the total imports of the seventeen North African and Middle Eastern countries listed in Table 6.

Because TIMTRADE is being developed in the first instance as an analytical tool for trade of the ECE countries (including their trade with countries in other regions), it is not yet adequate for tracing trade flows in other parts of the world, for example of sawn hardwood and plywood between Southeast Asia and the Middle East. This is apparent from a comparison of the third and fourth columns of Table 7 (recorded imports of thirteen North African and Middle Eastern countries in 1984 and import data for the same countries extracted from TIMTRADE, respectively). There are also cases, however (woodpulp), where the TIMTRADE data are higher than the recorded import totals or are unexplainably lower (paper and paperboard). This underlines the point already made about the reliability of much of the data on which this paper has had to be based.

TRADE IN SELECTED FOREST PRODUCTS BETWEEN THE ECE REGION AND NORTH AFRICA AND THE MIDDLE EAST

The rest of the paper will concentrate on the pattern of exports from Europe, the U.S.S.R., and North America to the five countries of North Africa and eight of the Middle East (Bahrain, Oman, Qatar, and Yemen excluded). As already seen in Table 7, these thirteen countries account for the bulk of imports into the two areas (100% in the case of North Africa) while imports of the other four are not included in the Middle Eastern eight for lack of data on origin.

The five products selected for trade flow analysis are sawn softwood, sawn hardwood, plywood, newsprint, and other paper and paperboard, the last group being a mixture of assortments of which printing and writing paper, other than newsprint, and packaging and wrapping paper and paperboard are the two main ones.

Table 8 shows the development of trade between the ECE countries and the thirteen North African and Middle Eastern countries between 1970 and 1984. The growth of trade in sawn

Table 7. Comparison of import data on forest products of North African and Middle Eastern countries with data on shipments to them of the exporting countries.

Product	Recorded* Imports of 17 Countries (from Table 6)		Recorded* Imports of 13 Countries	Import Data from TIMTRADE Data Base for 13 Countries**	
	Volume (1,000 units)	% of product group	Volume (1,000 units)	Total imports	Imports from ECE region
Sawnwood, total (m³)	7,386	100	--	--	--
Sawn softwood	6,546	89	6,537	5,938	5,745
Sawn hardwood	817	11	560	368	176
Sleepers	23	--	--	--	--
Wood-based panels, total (m³)	1,504	100	--	525	159
Plywood	1,242	83	1,116	452	99
Veneer sheets	--	--	--	24	12
Particleboard	--	--	--	29	29
Fiberboard	--	--	--	20	19
Woodpulp, total (m.t.)	188	100	188	259	259
Chemical	--	--	--	252	252
Mechanical and semichemical	--	--	--	7	7
Paper and paperboard, total (m.t.)	1,444	100	1,426	850	850
Newsprint	219	15	218	168	168
Printing and writing	370	26	360 ⎫	682	682
Other paper and paperboard	855	59	848 ⎭		
Sawlogs and veneer logs (m³)	945	100	945	306	306
Softwood	520	55	520	269	269
Hardwood	425	45	425	37	37

*FAO *Yearbook of Forest Products* 1985.
**The seventeen countries in Table 6 less Bahrain, Oman, Qatar, and Yemen Democratic Republic, which are not separately identified in the TIMTRADE data base.

softwood over the fourteen years is particularly striking in volume as well as percentage terms. The volume growth in the trade in paper and paperboard was also impressive. With regard to sawn hardwood, ECE countries shared to only a limited extent in the steep rise in imports of the thirteen shown in Table 9. While there may be some doubt as to the accuracy of the import data in Table 9, for reasons discussed earlier, there is no doubt that there has been a marked expansion in exports of these two products from the countries of Southeast Asia, mainly Malaysia, Singapore, the Philippines, Indonesia, and Taiwan to the Middle East. The figures in Table 5, taken from the 1985 *Yearbook of Forest Products,* give a partial indication of this.

Sawn Softwood

Table 10 places the ECE countries and the thirteen North African and Middle Eastern countries in the world context of trade in sawn softwood in 1984. Imports by the thirteen amounted to 5.94 million m³, or 8.3% of the world total. Of this virtually all, 5.75 million m³, or 8.0%, came from

Table 8. Trade between the ECE region (Europe, the U.S.S.R., North America) and thirteen Middle Eastern and North American countries,* in selected forest products.

| Product | Unit | Volume (1,000 units) | | | | Change (1970 to 1984) | |
		1970	1975	1980	1984	Volume	%
Sawn softwood	m^3	1,609	2,744	4,472	5,745	+4,136	+257
Sawn hardwood	m^3	128	251	183	176	+48	+38
Plywood	m^3	37	59	94	99	+62	+168
Newsprint	m.t.	88	90	129	168	+80	+91
Other paper and paperboard	m.t.	384	582	607	682	+298	+78

*As shown in Table 11.

Table 9. Total imports of thirteen Middle Eastern and North African countries,* reported by the importing countries.

| Product | Unit | Volume (1,000 units) | | | | Change (1970 to 1984) | |
		1970	1975	1980	1984	Volume	%
Sawn softwood	m^3	2,070	2,653	4,145	6,537	+4,467	+216
Sawn hardwood	m^3	126	445	765	560	+434	+344
Plywood	m^3	90	253	639	1,116	+1,026	+1,140
Newsprint	m.t.	111	139	155	218	+107	+96
Other paper and paperboard	m.t.	505	718	961	1,208	+703	+139

*As shown in Table 11.

ECE countries. European exporters' share was 4.47 million m^3 (6.2%), North America's 928,000 m^3 (1.3%), and the U.S.S.R.'s 347,000 m^3 (0.5%). The importance of the thirteen overseas markets to the exporters was 19.6% for Europe on average (see Table 12 for country details), 2.3% for North America, and 4.9% for the U.S.S.R.

The largest exporters to the thirteen (Table 11) in 1984 were Sweden (1.41 million m^3), Finland (1.38 million m^3), Canada (869,000 m^3), and Austria (634,000 m^3). The importance of these markets to the exporters varied from 29% for Finland to 2% for Canada. However, for some of the smaller exporters, the thirteen countries accounted for the major part of their total shipments: Yugoslavia (76%), Spain (70%), and Romania (an estimated 61%); geographic proximity and established commercial links no doubt play an important role.

The largest country-to-country trade flow in sawn softwood in 1984 was from Sweden to Egypt: 675,000 m^3 (8.4% of Sweden's total exports and 33% of Egypt's imports). This was followed by Finland's exports to Algeria and Egypt (473,000 m^3 and 454,000 m^3, respectively) and Sweden's exports to Saudi Arabia (317,000 m^3). Saudi Arabia was an important purchaser from most of the other ECE countries, notably Canada, Austria, and Romania; while Egypt and Algeria were also large buyers from Austria, the U.S.S.R., and Canada; and Egypt was from Romania.

It was not possible in preparing this paper to look into the reasons for particular trade patterns,

Table 10. Pattern of trade in sawn softwood between regional groups in 1984.

Importers	World total	Europe	U.S.S.R.	North America	ECE region total	Rest of world†
			Exporters			
			$(1,000\ m^3)$			
World	71,788	22,836	7,094	40,879	70,809	979
Europe	26,157	17,952	5,608	2,479	26,039	118
North America	31,990	--	--	31,968	31,968	22
5 North African*	3,637	2,588	326	571	3,485	152
8 Middle Eastern**	2,301	1,882	21	357	2,260	41
Subtotal (13 countries)	5,938	4,470	347	928	5,745	193
Rest of world†	7,703	414	1,139	5,504	7,057	646
			(% of world total)			
World	100.0	31.8	9.9	56.9	98.6	1.4
Europe	36.4	25.0	7.8	3.5	36.3	0.2
North America	44.6	--	--	44.5	44.5	--
5 North African	5.1	3.6	0.5	0.8	4.9	0.2
8 Middle Eastern	3.2	2.6	--	0.5	3.1	--
Subtotal (13 countries)	8.3	6.2	0.5	1.3	8.0	0.3
Rest of world	10.7	0.6	1.6	7.7	9.8	0.9

Source: FAO/ECE, TIMTRADE data base.
Note: Detail may not add to total, because of rounding.

*Algeria, Egypt, Libya, Morocco, Tunisia.
**Iran, Iraq, Israel, Jordan, Kuwait, Lebanon, Saudi Arabia, Syria.
†Including unidentified trade flows.

although it would be of considerable interest to do so. Why, for example, did Sweden and Yugoslavia export so much to Egypt, but almost nothing to Algeria, when Finland and some other countries sold large quantities to both?

Sawn Hardwood

The volume of exports of sawn hardwood that can be readily identified from ECE countries to the thirteen countries amounted to 176,000 m^3 in 1984 (Table 12). There is reason to believe that the volume may be somewhat higher, because it does not include any exports to those countries from Romania, which is not at present reporting trade flow data. Even taking into account possible values from Romania, ECE exports, virtually all from Europe, accounted for only about one-third of the thirteen's total imports of sawn hardwood, estimated at over 500,000 m^3. Of the remaining two-thirds, more than half can be traced to the trade between Singapore, Malaysia, and the Philippines as exporters and Saudi Arabia and Oman as importers (Table 13).

Table 11. Trade in sawn softwood between selected countries in 1984.

	Exporters										10 countries		All Other*	Total
Importers	Austria	Finland	Portugal	Romania	Spain	Sweden	Yugoslavia	U.S.S.R.	Canada	United States	Volume	% of total		
						(1,000 m³)								
Algeria	96	473	--	--	--	1	3	78	441	--	1,092	100	--	1,092
Egypt	122	454	--	211	11	675	60	248	99	7	1,887	92	155	2,042
Libya	--	1	--	--	1	21	5	--	--	1	29	45	35	64
Morocco	--	3	37	--	170	41	20	--	22	1	294	97	9	303
Tunisia	--	74	2	--	8	12	17	--	--	--	113	83	23	136
Iran	119	38	--	--	--	112	--	--	--	--	269	97	8	277
Iraq	10	153	--	--	1	164	6	--	--	--	334	99	2	336
Israel	2	109	30	--	2	26	44	--	17	10	240	96	11	251
Jordan	--	--	--	--	--	29	--	--	--	1	30	100	--	30
Kuwait	9	--	--	--	--	--	7	--	--	--	16	94	1	17
Lebanon	12	--	--	--	--	2	5	--	--	15	34	100	--	34
Saudi Arabia	260	58	1	256	--	317	26	21	290	24	1,253	96	49	1,302
Syria	4	20	--	--	--	10	20	--	--	--	54	100	--	54
13 countries	634	1,383	70	467*	193	1,410	213	347	869	59	5,645	95	293	5,938
% of total	16%	29%	6%	61%	70%	18%	76%	5%	2%	2%	8%	--	6%	8%
All other	3,329	3,420	1,084	301*	81	6,601	69	6,747	36,262	3,689	61,583	94	4,267	65,850
Total	3,963	4,803	1,154	768	274	8,011	282	7,094	37,131	3,748	67,228	94	4,560	71,788

Source: FAO/ECE, TIMTRADE data base.
*Including ECE countries not shown in table.

Table 12. Pattern of trade in sawn hardwood between regional groups in 1984.

Importers	World total	Europe	North America	ECE region total*	Rest of world†
			Exporters		
			$(1,000 \text{ m}^3)$		
World	13,688	3,001	1,373	4,501	9,187
Europe	5,449	2,444	366	2,815	2,634
North America	1,407	3	627	630	777
5 North African**	276	156	--	156	120
8 Middle Eastern**	260	14	5	19	241
Subtotal (13 countries)	536	170	5	176	360
Rest of world*†	6,296	384	375	880	5,416
			(% of world trade)		
World	100.0	21.9	10.0	32.9	67.1
Europe	39.8	17.9	2.7	20.6	19.2
North America	10.3	--	4.6	4.6	5.7
5 North African**	2.0	1.1	--	1.1	0.9
8 Middle Eastern**	1.9	0.1	--	0.1	1.8
Subtotal (13 countries)	3.9	1.2	--	1.3	2.6
Rest of world*†	46.0	2.8	2.7	6.4	39.6

*Including U.S.S.R.
**As in Table 10.
†Including unidentified type trade flows.

Of the European exporters, only Yugoslavia ships significant quantities of sawn hardwood to the thirteen; in 1984 this trade amounted to 140,000 m³. However, of this volume, 125,000 m³ went to just one market—Egypt. France and possibly Romania are the only other countries with moderate volumes of exports to these markets. Recalling the trends since 1970 as shown earlier in Tables 8 and 9, it appears that European and North American exporters, apart from Yugoslavia, have not been able, or have not really attempted, to hold their share of the North African and Middle Eastern markets in face of the strong competition from other areas, notably Southeast Asia.

Plywood

A rather similar situation exists for plywood. ECE countries' exports to the thirteen in 1984 amounted to 99,000 m³ (Table 14), which was 22% of their estimated total imports of 452,000 m³. Even allowing for the approximate nature of the import data, it can be seen from Table 9 that plywood imports have risen rapidly into North Africa and the Middle East, especially the latter, but that Europe has not been able to benefit very much—and even less North America. Again, it has been the countries of Southeast Asia, as well as Taiwan and South Korea, that have greatly

Table 13. Trade in sawn hardwood and plywood between selected East and Southeast Asian exporters and North African and Middle Eastern importers in 1984.

	Exporters						
Importers	Taiwan	South Korea	Indonesia	Malaysia	Phil- ippines	Singapore	Total (countries shown)
				$(1,000 \text{ m}^3)$			
Sawn hardwood							
Oman	--	--	--	29	13	9	51
Saudi Arabia	--	--	--	63	7	95	165
Total (2 countries)	--	--	--	92	20	104	216
Plywood							
Egypt	20	10	36	--	--	5	71
Libya	--	16	--	--	--	2	18
Kuwait	13	1	--	3	--	51	68
Saudi Arabia	36	35	--	3	--	87	161
Total (4 countries)	69	62	36	6	--	145	318

Source: FAO *Yearbook of Forest Products* 1985.

expanded their trade. Some trade flow details for 1984 are given in Table 13; exports from Singapore to Saudi Arabia and Kuwait were the largest in volume among identified trade flows. Further analysis of the trade data is needed to determine the origin of the plywood supposedly imported into the region; or to check whether the estimates of total imports in Table 6 are in fact reasonable or possibly on the high side.

Only a few European countries exported more than 10,000 m^3 to the thirteen in 1984: Romania (26,000 m^3 or possibly more), Yugoslavia (25,000 m^3), Belgium and Greece (11,000 m^3 each). Shipments to the thirteen of the three largest exporters of the ECE region—Finland, Canada, and the U.S.S.R.—amounted in aggregate to only 9,000 m^3, a surprisingly low figure, especially when compared with their sawn softwood exports.

Paper and Paperboard

As was true for plywood, there is some reason to doubt whether the total estimated imports of newsprint and other paper and paperboard shown in Table 6 are realistic. The ECE countries account for the major part of world exports of these products, and their exports to the thirteen countries in 1984 (Tables 15 and 16) were substantially less than the import estimates. According to TIMTRADE data, the thirteen imported 168,000 tons of newsprint and 682,000 tons of other paper and paperboard from Europe, the U.S.S.R., and North America in 1984 and no volumes from other sources. That zero figure may not be quite accurate, but the volumes coming from other sources were likely to be quite small.

As shown in Table 17, Egypt, Israel, and Iran were the largest importers of newsprint, as well as of other categories of paper. Sweden was the largest supplier of paper in total, followed by the United States, Finland, and Canada. Other countries with substantial exports to the thirteen in 1984 were Spain, the Federal Republic of Germany, Yugoslavia, and Italy.

Table 14. Pattern of trade in plywood between regional groups in 1984.

Importers	World total	Europe	U.S.S.R.	North America	ECE region total	Rest of world*
			Exporters			
			$(1,000 \text{ m}^3)$			
World	8,416	1,503	380	715	2,598	5,818
Europe	3,029	1,262	300	553	2,115	914
North America	1,412	23	12	78	113	1,299
5 North African	166	76	--	--	76	90
8 Middle Eastern	287	23	--	--	23	264
Subtotal (13 countries)	452	99	--	--	99	354
Rest of world*	3,523	119	68	84	271	3,252
			(% of world total)			
World	100.0	17.9	4.5	8.5	30.9	69.1
Europe	25.5	15.0	3.6	6.6	25.1	10.9
North America	11.9	0.3	0.1	0.9	1.3	15.4
5 North African	1.4	0.9	--	--	0.9	1.1
8 Middle Eastern	2.4	0.3	--	--	0.3	3.1
Subtotal (13 countries)	3.8	1.2	--	--	1.2	4.2
Rest of world	44.8	1.4	0.8	1.0	3.2	38.6

Source: FAO/ECE, TIMTRADE data base.
Note: Detail may not add to total, because of rounding.
*Including unidentified trade flows.

Table 15. Pattern of trade in newsprint between regional groups in 1984.

Importers	World total	Europe	U.S.S.R.	North America	ECE region total	Rest of world*
			Exporters			
	(1,000 m.t.)					
World	13,272	4,015	341	8,411	12,767	505
Europe	3,349	2,966	218	160	3,344	5
North America	7,029	337	--	6,692	7,029	--
5 North African	103	62	--	41	103	--
8 Middle Eastern	65	65	--	--	65	--
Subtotal (13 countries)	168	127	--	41	168	--
Rest of world*	2,726	585	123	1,518	2,226	500
	(% of world total)					
World	100.0	30.2	2.6	63.4	96.2	--
Europe	25.2	22.3	1.6	1.2	25.2	--
North America	53.0	2.5	--	50.4	--	--
5 North African	0.8	0.5	--	0.3	0.8	--
8 Middle Eastern	0.5	0.5	--	--	0.5	--
Subtotal (13 countries)	1.3	1.0	--	0.3	1.3	--
Rest of world	20.5	4.4	0.9	11.4	16.8	3.8

Source: FAO/ECE, TIMTRADE data base.
Note: Detail may not add to total, because of rounding.
*Including unidentified trade flows.

Middle East and North Africa 69

Table 16. Pattern of trade in paper and paperboard other than newsprint between regional groups in 1984.

			Exporters			
Importers	World total	Europe	U.S.S.R.	North America	ECE region total	Rest of world*
			(1,000 m.t.)			
World	26,312	18,595	645	4,917	24,157	2,155
Europe	15,885	14,207	492	947	15,646	239
North America	2,974	1,011	--	1,821	2,832	142
5 North African	301	243	--	58	301	--
8 Middle Eastern	382	234	--	148	382	--
Subtotal (13 countries)	682	477	--	205	682	--
Rest of world*	6,771	2,900	153	1,944	4,997	1,774
			(% of world total)			
World	100.0	70.7	2.5	18.7	91.8	8.2
Europe	60.4	54.0	1.9	3.6	59.5	0.9
North America	11.3	3.8	--	6.9	10.8	0.5
5 North African	1.1	0.9	--	0.2	1.1	--
8 Middle Eastern	1.5	0.9	--	0.6	1.5	--
Subtotal (13 countries)	2.6	1.8	--	0.8	2.6	--
Rest of world	25.7	11.0	0.6	7.4	19.0	6.7

Source: FAO/ECE, TIMTRADE data base.
Note: Detail may not add to total, because of rounding.
*Including unidentified trade flows.

Table 17. Trade in newsprint and other paper and paperboard between selected countries in 1984.

| Importers | Exporters | | | | | |
| | Newsprint | | | Other paper and paperboard | | |
	Finland	Sweden	Canada	Finland	Sweden	U.S.
	(1,000 m.t.)					
Algeria	--	20.0	--	6.0	30.4	--
Egypt	20.8	4.8	41.1	23.2	18.4	40.8
Libya	--	--	--	--	--	--
Morocco	4.9	1.0	--	10.4	35.8	1.1
Tunisia	--	1.9	--	5.8	6.2	7.3
North Africa (5 countries)	25.7	27.7	41.1	45.4	90.8	49.2
Iran	23.2	0.3	--	47.0	43.9	--
Iraq	--	--	--	--	--	--
Israel	3.3	26.7	--	15.3	9.6	68.6
Jordan	--	--	--	--	--	4.3
Kuwait	--	--	--	--	--	1.9
Lebanon	5.3	--	--	10.2	6.7	23.2
Saudi Arabia	--	--	--	--	--	37.2
Syria	--	--	--	--	--	--
Middle East (8 countries)	31.8	27.0	--	72.5	60.2	135.2
Total (13 countries)	57.5	54.7	41.1	117.9	151.0	184.4

DEVELOPMENTS SINCE 1984

The impression left by a review of trends up to 1984 is of an impressive upward growth in imports of forest products into North Africa and the Middle East. More recent developments have shown how misleading that impression could be.

What has happened, of course, has been the steep slide during the latter part of 1985 and 1986 in oil prices. This process had, in fact, begun in 1982, but between 1982 and 1985 the drop had been gradual, as may be seen in Figure 2. Largely as a consequence, the current account surpluses of the oil-exporting countries, which had been massive in the late 1970s and early 1980s, were reduced to zero or even negative values in the following years. For example, the combined current account surpluses of Bahrain, Kuwait, Oman, Qatar, Saudi Arabia, and the United Arab Emirates in 1980 and 1981 stood at around U.S.$70 billion, fell to $20 billion in 1982, were about $5 billion in deficit in 1983 and 1984, about in balance in 1985, but there was again a deficit of an estimated $13 billion in 1986 (*Barclay's Review*, February 1987).

The oil-exporting countries' potential for importing goods and services was thus seriously curtailed. In addition, it seems that in several of them the massive phase of construction during the 1970s and early 1980s has now been completed and, at least as far as sawnwood and panels are concerned, import requirements are unlikely to recover in the foreseeable future, if ever, to the all-time peaks reached, for many of the countries concerned, in 1984. Data for 1986 are just beginning to come in, and those of the three major European sawn softwood exporting countries—Austria,

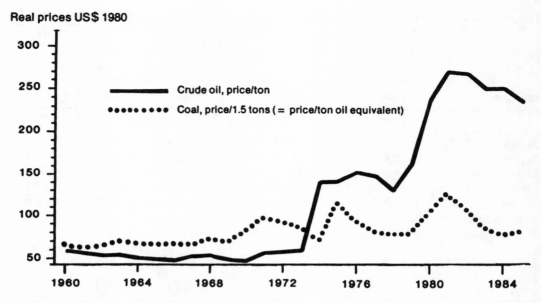

Figure 2a. Import prices (c.i.f.) for crude oil and steam coal to Sweden, 1960-84. Source: Radetzki (1986).

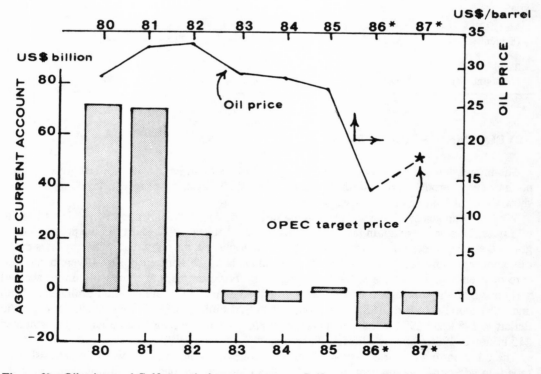

Figure 2b. Oil prices and Gulf countries' current account. Gulf countries: Bahrain, Kuwait, Oman, Qatar, Saudi Arabia, and United Arab Emirates. Sources: IMF, national sources, and Barclay's estimates and forecasts.

*Estimate for 1986; forecast for 1987.

Finland, and Sweden—are shown in Table 18 and Figure 3. In aggregate, their exports to their main North African and Middle Eastern clients fell from a peak 3.46 million m³ in 1984 to 1.88 million m³ in 1986, most of the fall occurring between 1985 and 1986.

An interesting aspect has been the influence of this trade flow on the three exporting countries' total exports over the past decade or so. The share of their exports to North Africa and the Middle East rose as total exports rose, from 2.2% of total exports in 1974 to 19.0% in 1984, before falling back to 11.9% in 1986 (Figure 4). Their exports to all other markets over the twelve years have been, with the exception of 1979 when they were appreciably higher, within the range of 13 to 15 million m³ a year with no obvious upward or downward trend.

The drop in imports in 1985-86 was no doubt accentuated by a destocking movement by importers and a temporary cessation of buying. Imports in 1987 could show some recovery, also helped by the efforts by OPEC countries to underpin the price of oil at $18 a barrel. Nevertheless, 1984 could prove to have been the peak year for imports of sawnwood and wood-based panels into North Africa and the Middle East, unless a new series of remarkable events occur. One development that would give some stimulation to imports would be an ending of the Iran-Iraq war, which would be followed by a period of reconstruction in those countries. Another would be the ending of the internal conflicts in Lebanon.

Statistics for 1986 for trade in paper and paperboard are not yet to hand. Some reduction could be anticipated, compared with 1984 and 1985, but probably not on the same scale as for sawn softwood. In the longer term, requirements for paper and paperboard are likely to expand steadily in line with growth of the countries' economies, populations, and standards of living.

DISCUSSION AND CONCLUSION

This paper has attempted, using a somewhat inadequate statistical base, to trace the development of imports of forest products into North Africa and the Middle East over the past decade and a half. Growth in this trade has been explosive, at least up to 1984, and to a large extent fueled by the expanding revenues of the oil-exporting countries. Developments since 1984, and indeed those during the 1970s, have demonstrated the volatility of the economic and monetary situation of those countries. There is no basis, therefore, on which to build a reasonable scenario of future events, either in general or more specifically related to trade in forest products.

The most reasonable—and obvious—statement would be that import demand in North Africa and the Middle East will be related in some way to the development of oil prices, coupled with the share of total world demand for oil that is provided by the OPEC countries in the region. Quite a number of qualifications need to be made, however, among which are the following.

1. By no means are all the countries important oil producers and exporters. Some, including Egypt, Morocco, Tunisia, and Israel, have other income-generating sectors, such as tourism.

2. Massive construction programs have been carried out in several of the OPEC countries, which drew in large quantities of imports. Many of these programs have been completed or are approaching completion, and, as happened in Europe in the mid-1970s, construction activity may fall back before stabilizing at a level below former peak levels.

3. The countries are, because of their own lack of forest resources, largely without a tradition in the use of wood products. This has a number of implications: (a) The relation between specification and suitability for a particular end use may not be fully appreciated. There is a tendency to take price as the basic criterion, regardless of the quality needed to fulfill a specific use. (b) As a corollary of (a), a concerted effort of technical promotion by the exporters could bring positive results, not only in terms of quantity bought and used in the North Africa and Middle Eastern countries but also in a better appreciation of the quality of wood products and a greater willingness to use quality as well as price as a purchasing criterion. (c) The markets are individual, and success in one is no guarantee of equal success in another. There is no substitute for gaining a sound

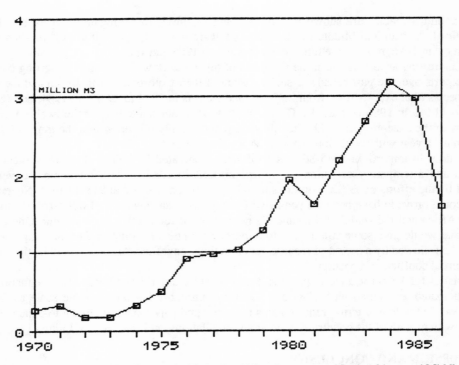

Figure 3. Sawn softwood exports by Austria, Finland, and Sweden to seven North African and Middle Eastern countries, 1970-86.

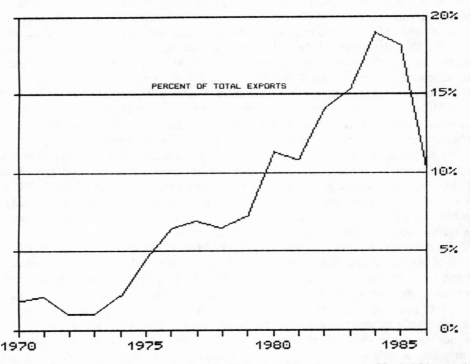

Figure 4. Share of sawn softwood exports by Austria, Finland, and Sweden to seven North African and Middle Eastern countries in total exports, 1970-86.

Table 18. Exports of Austria, Finland, and Sweden to North Africa and the Middle East, 1984 to 1986 (1,000 m³).

	Exporters											
	Austria*			Finland**			Sweden†			Total (3 countries)		
Importers	1984	1985	1986	1984	1985	1986	1984	1985	1986	1984	1985	1986
Algeria	95.8	83.5	169.5	473.0	395.5	133.3	1.2	36.0	0.1	570.0	515.0	302.9
Egypt	21.9	23.9	9.1	454.4	584.0	356.9	675.3	780.4	548.7	1,151.6	1,388.3	914.7
Libya	121.9	12.2	79.4	0.7	0.6	0.2	21.4	17.2	14.6	144.0	30.0	94.2
Morocco	--	--	--	2.9	1.8	1.3	41.0	58.2	52.1	43.9	60.3	53.4
Tunisia	0.4	0.2	0.7	74.3	40.6	27.5	11.5	29.3	--	86.2	70.1	28.2
Subtotal	240.0	119.8	258.7	1,005.3	1,022.5	519.2	750.4	921.4	615.5	1,995.7	2,063.7	1,393.4
Iran	118.8	95.2	69.7	38.0	21.1	7.8	111.5	--	--	268.3	116.3	77.5
Iraq	10.2	67.9	1.2	152.5	180.3	--	164.0	151.7	2.2	326.7	399.9	3.4
Israel	2.2	4.6	‡	109.0	96.4	124.5	26.2	25.5	32.0	137.4	126.5	156.5
Jordan	0.3	8.9	--	--	--	--	28.5	36.6	25.6	28.8	45.5	25.6
Kuwait	8.9	12.8	6.8	--	--	--	--	--	--	8.9	12.8	6.8
Lebanon	12.1	2.1	--	--	--	--	2.3	--	--	14.4	2.1	--
Saudi Arabia	259.8	174.3	37.9	58.3	65.9	51.0	317.3	183.7	124.1	635.4	423.9	213.0
Syria	3.8	4.0	--	19.9	0.2	--	10.2	--	--	33.9	4.2	--
Other Middle East	6.2	18.9	6.9	--	0.3	--	--	--	--	6.2	19.2	6.9
Subtotal	422.3	388.7	122.5	377.7	364.3	183.3	660.0	397.5	183.9	1,460.0	1,150.4	489.7
Total listed countries	662.3	508.5	381.2	1,383.0	1,386.8	702.5	1,410.4	1,318.9	799.4	3,455.7	3,214.1	1,883.1
% of total exports	16.7%	14.2%	10.7%	28.8%	28.4%	15.5%	17.6%	16.8%	10.3%	20.6%	19.7%	11.9%
Total exports	3,962.5	3,592.2	3,546.2	4,805.2	4,884.3	4,543.7	8,011.2	7,865.7	7,731.9	16,778.9	16,342.2	15,821.8

*Bundesholzwirtschaftsrat.
**Finnish Sawmill Owners' Association.
†Swedish Wood Exporters' Association and FAO/ECE Timber Bulletin.
‡Included in "other Middle East."

Middle East and North Africa 75

understanding, from representatives or agents on the spot, of how the markets work. (*d*) One market aspect that may grow in importance in the North Africa and Middle Eastern area as elsewhere is the use of countertrading (barter), which may require trading expertise not directly available in wood exporting companies and thus require linkups with trading companies with such experience. (*e*) Exporters must be flexible enough to cope with sharp fluctuations in purchasing by North African and Middle Eastern importers, sometimes caused by factors external to the trade itself (political, economic, monetary).

The markets of North Africa and the Middle East will probably continue to be "difficult" ones for forest products exporters, especially those from countries with a different social and cultural background, who may find problems in understanding and adjusting to different ways of doing business. Nonetheless, in aggregate, these markets will probably retain the 5% or so of total world import value of forest products to which they had grown during the 1970s and will be rewarding to those who are prepared to take the time and trouble to cultivate them.

One point is sure: the countries covered in this paper will always be dependent on imports for all but a small fraction of their total requirements of forest products. With populations likely to continue to grow at above world average rates, their latent needs for shelter (sawnwood and panels) and education and culture (paper) will expand. To what extent these needs can be met will depend to a large extent on the health of their economies and trade balances. For many years, the region has been disturbed by conflicts, which have in many countries delayed economic and social development. One might speculate on what would be the impact on import demand for all types of forest products if peace should one day come to all parts of the region.

REFERENCES

FAO. 1984-85. Yearbook of forest products. Food and Agriculture Organization of the United Nations, Rome.

———. 1985. World forest resources 1980. FAO, Rome. 22 p.

FAO/ECE. Annual. Timber Bulletin for Europe. (Timber Bulletin, since 1985).

———. 1976. European timber trends and prospects, 1950 to 2000. United Nations, New York. 308 p.

———. 1986. European timber trends and prospects to the year 2000 and beyond. United Nations, New York. 594 p. (2 vols.)

International economy: Developing countries. Barclay's Review 62, no. 1 (February 1987):23-27.

Radetzki, M. 1986. Outlook for oil and coal prices in international trade. Skandinaviska Enskilda Banken Quarterly Review (Stockholm) 4:98-108.

Thirgood, J. V. 1981. Man and the Mediterranean forest. Academic Press, London. 194 p.

UNCTAD/GATT. 1980. Survey of selected Middle Eastern markets for wood-based building materials from developing countries. International Trade Centre, Geneva. 154 p.

UNDP. 1985. 1985—And towards the 1990s. Annual report of the United Nations Development Programme, New York.

Chapter 6

An Econometric Analysis of Supply and Demand for Forest Products in Taiwan

SUN JOSEPH CHANG

Rapid economic growth since the late 1950s and early 1960s has propelled Taiwan from a developing country to what is now called a newly industrialized country (see data in Appendix 1). With an export-oriented economy and a rather small land base (36,000 km^2), Taiwan relies ever more heavily on foreign countries as both the source of raw material and the market for finished products (thus the adage, "Use foreign resources to earn foreign exchange"). Forest products industry, as an important sector of the economy, is no exception. As shown in Appendix 1, the import share of logs for solid-wood consumption rose from 16.1% in 1961 to 89.46% in 1984. As Taiwan's economy continues to grow, its heavy reliance on imported raw material for the wood products industry should continue and its status as an important market should be enhanced. Since Taiwan is a major user of wood attempting to expand its exports of wood products, it is clearly in the best interest of the United States to have a better understanding of this important market.

This paper will present an econometric analysis of the supply and demand for solid-wood products in Taiwan (Chang and Jen 1986). To facilitate the analysis and the presentation, the model for solid-wood products will be divided into four parts: the plywood industry submodel, the furniture industry submodel, the softwood submodel, and the hardwood submodel. The plywood industry submodel includes equations for production, domestic consumption, domestic price index, export, export price, and import. The furniture industry submodel includes equations for furniture production and export. The softwood submodel includes equations for softwood log supply, demand, price, export, and import. The hardwood submodel includes equations for hardwood log supply, demand, import, and price. In addition, the hardwood submodel includes an equation for hardwood lumber import. All together, these equations made up a simultaneous equations system to describe the supply and demand situation for solid-wood products in Taiwan.

THE SOLID-WOOD PRODUCTS MODEL FOR TAIWAN

The specific functional forms of all eighteen equations and a list of variables are shown in Appendix 2. Before discussing any particular equation, it should be pointed out that generalized Cobb-Douglas production functions involving capital, labor, energy, and raw material were assumed throughout the model to derive either an output supply function or an input factor demand function for raw material. (See Appendix 3.)

Equation 3 for the production of plywood in Appendix 2 is almost exactly identical to equation 6 in Appendix 3 with the appropriate variables replacing P, PK, PL, PE, and PM respectively. In addition, a trend variable 1/(YR-1952) is added to allow for variable rates of technical progress. On the other hand, the factor input demand for imported hardwood logs for the plywood and furniture industries combined is shown as equation 17 in Appendix 2.

Consumption function is expressed in a simple antilog function of the type

$$Q = a \, (GNP)^b P^c$$

where Q represents the quantity of output consumed and a, b, and c are constants. GNP represents gross national product and P is the price of output. As such, coefficient b would represent the income elasticity, while c would represent the price elasticity. Thus the basic model is modified for the plywood consumption function so that equation 4 of Appendix 2 is used in the model. Essentially the parameter AD2 represents the income effect in terms of housing starts and the parameter AD3 represents the price effect of domestic price index for plywood. In the consumption for softwood logs, the price of a close substitute, the price of imported hardwood logs (PRMHLG), is added to the model to incorporate the possible effect of substitution.

Export equations consist of two types: (1) when exporters are price setters, including furniture export and softwood log export, and (2) when exporters can affect both price and quantity of export. For this type, only the plywood export fits the description. When exporters are price setters, the export quantity is basically a derived demand of the importing country or countries. In this sense, exports of furniture are basically dictated by U.S. demand and softwood log exports are basically dictated by Japanese housing starts. On the other hand, the export of plywood is determined by the production of plywood and the price the plywood industry can get in the export market. Furthermore, the export of plywood also involves the formation of the export price.

Price formation, including the formation of export price for plywood, is basically a function of the previous year's price and either the consumption or the export of that particular commodity of the same year; for example, see the equation for the export price of plywood (equation 7, Appendix 2). Basically, a function of this type provides partial adjustments of the price, and the coefficient for the previous year's price indicates the percentage (in decimal) of price adjustment.

Together the eighteen equations form a nonlinear simultaneous equation system to describe the supply and demand for solid wood in Taiwan. Presented below are the results from the two-stage least squares (2SLS) analysis of the data from 1961 to 1984. An SAS computer package was used to solve these equations (SAS Institute Inc. 1984).

THE WOOD FURNITURE INDUSTRY SUBSECTOR

The wood furniture industry in Taiwan is an important export-oriented industry in Taiwan's economy. In order to fully understand this industry, it is imperative to first understand the export of wood furniture.

Export Value of Wood Furniture (EXFUR) (Equation 1)

The United States is the most important export market for the wood furniture industry in Taiwan. Indeed, Taiwan has now emerged as the number one foreign supplier of wood furniture to the United States. Thus real GNP in the United States (RGNPUSA) is chosen as the explanatory variable of export market. As shown in Table 1, the relationship between the export of wood furniture and real GNP of the United States is highly significant (AA1 and AA2). Furthermore, since the export value of wood furniture and the real GNP of the United States display an S-shaped relationship, as the GNP increases the wood furniture export value would also increase, yet at a decreasing rate. Based on the 1984 real GNP for the United States, a 1% increase in real GNP would result in a 7.8% increase in this furniture export. It is no small wonder then that domestic producers are feeling the pressure from overseas competition.

Table 1. Parameters associated with equations of furniture and plywood industries.

Equation	Parameter	Estimate	Approximate Standard Error	t Value
1	AA1	13.71	0.83	16.44**
	AA2	−12.21	1.30	−9.41**
2	AB1	3.15	3.99	0.79
	AB2	0.56	0.26	2.14*
	AB3	−0.10	0.14	−0.74
	AB4	0.62	0.47	1.32
	AB5	−0.36	0.16	−2.17*
	AB6	−0.43	0.21	−2.03
3	AC1	19.59	5.15	3.81**
	AC2	1.11	0.91	1.22
	AC3	−0.32	0.27	−1.17
	AC4	−0.41	0.43	−0.95
	AC5	−0.42	0.27	−1.55
	AC6	−0.28	0.51	−0.54
	AC7	−55.40	25.47	−2.17*
4	AD1	6.82	1.08	6.30**
	AD2	0.36	0.21	1.77
	AD3	0.31	0.52	0.59
	AD4	0.04	0.02	1.80
5	AE1	−0.19	1.14	−0.17
	AE2	0.71	0.21	3.34**
	AE3	0.10	0.13	0.74
	AE4	0.02	0.14	0.13
6	AF1	0.50	0.65	0.77
	AF2	−0.20	0.03	−6.92**
	AF3	1.02	0.05	19.82**
7	AG1	−1.32	1.03	−1.29
	AG2	0.89	0.06	13.91**
	AG3	0.15	0.08	1.86

* 5% significance.   ** 1% significance.

Value of Wood Furniture Production (PROFUR) (Equation 2)

Conceptually, furniture production is no different from any other industry. Empirically, the furniture industry does pose some difficulties. Because of the wide variety of furniture items, total quantity of furniture produced in terms of adding, for example, tables and chairs provides very little empirical meaning. In order to regress production value as the dependent variable for the furniture industry, one must also come up with a proxy for the relatively meaningless average price per piece of furniture. As a proxy, the export value of wood furniture was chosen under the belief that total value of furniture export provides some indication of the price the furniture producers are able to command for a piece of furniture. As shown in Table 1, when EXFUR increases 1%, PROFUR increases 0.56% (AB2 = 0.56). In addition, results from the analysis shown in Table 1 also suggest:

A 1% change in	results in a change of PROFUR of
Real black market rate (RBMR)	−0.10% (AB3 = −0.10)
Wage rate in wood products industry (WRWPI)	0.62% (AB4 = 0.62)
Price of electricity (PRELEC)	−0.36% (AB5 = −0.36)
Price of imported hardwood logs (PRMHLG)	−0.43% (AB6 = −0.43)

Among the costs of factor inputs, only the coefficient for the price of electricity is significantly different from zero. That is to say, changes in the cost of the other three factors do not significantly affect the production value of the furniture industry. In terms of the sign of the coefficients, all but the wage rate have the right sign. While the result may deviate from what the theory would suggest, it actually reflects the speed of wage rate adjustment during the business cycle. When the industry is doing well, the wage rate goes up quickly. When the industry is not doing well, the wage rate goes down very quickly. Consequently, the result of the analysis suggests the positive correlation between wage rate and production value.

THE PLYWOOD INDUSTRY SUBSECTOR

The plywood industry has been a major exporting industry in Taiwan. It started importing the Philippine lauan in 1949 to manufacture hardwood plywood. Ever since, it has been almost exclusively a hardwood plywood industry based on hardwood peeler logs from Southeast Asia. With prudent management, lower labor cost, and ever-improving manufacturing technology, the plywood industry soon won a worldwide reputation as a major supplier of hardwood plywood. By 1973, the plywood industry ranked behind the textile and electronic industries as the third largest exporting industry. In recent years, however, the plywood industry in Taiwan has fallen on hard times. The countries in Southeast Asia that supply peeler logs—mainly Indonesia—in an attempt to build up a domestic plywood industry have first raised the price of export logs and then banned such exports entirely. The data in Appendix 1 show the early growth and eventual decline of the plywood industry from 1961 to 1984.

The Production of Plywood (Equation 3)

The production of plywood depends on the price index of plywood (PRIDPLY), real black market interest rate (RBMR), wage rate in wood products industry (WRWPI), price of industrial fuel oil (PRFUEL), and price of imported hardwood logs (PRMHLG). The results shown in Table 1 indicate that when the price index of plywood increases 1%, plywood production increases 1.11% (AC2). Since the coefficient for the price index equals $s/(1-s)$, it must be true that $s<1$. The plywood industry, therefore, displays decreasing return to scale. The bigger the scale of production, the higher the cost of production. The coefficients for the four factors of production are all negative and agree with the result of theoretical derivation. Thus when the costs of production factors increase and the price index for plywood stays constant, the production of plywood decreases. The coefficient for the technical trend is –55 (AC7), suggesting that the technology is still improving.

The Consumption of Plywood (Equation 4)

Plywood consumption is affected by amount of housing starts expressed in square meters, price index of plywood, and time trend. For every 1% increase in area of housing starts, consumption of plywood increases 0.36% (AD2). When plywood price index increases 1%, plywood consumption also increases 0.31% (AD3). This result is contrary to what the theory would suggest. Statistically, however, the result is not significantly different from zero, suggesting that the effect of price on plywood consumption is rather small. The trend variable indicates that every year the consumption of plywood is increasing at a rate of about 4% (AD4 = 0.04).

The Price Index of Plywood (Equation 5)

Plywood price index is a function of the previous year's price, plywood consumption, and plywood export. Where the previous year's price index of plywood increases 1%, the current year's price index increases 0.71% (AE2). When the consumption of plywood increases 1%, price index for plywood increases 0.10% (AE3). A 1% increase in plywood export leads to a 0.02% increase in the price index. These latter two effects, however, are both statistically insignificant.

Plywood Export (Equation 6)

Plywood export is mainly affected by the export price of plywood and the production of plywood. When the price of export plywood increases 1%, plywood export decreases 0.20% (AF2), and the coefficient is statistically very significant. This result lends credibility to the report that there used to be intense price cutting among plywood mills for the export market. In addition, higher production of plywood also leads to higher export volume; namely, a 1% increase in production leads to a 1.02% (AF3) increase in export of plywood. Therefore, there should be no doubt then that the plywood industry has always been an important export-oriented industry.

Export Price of Plywood (Equation 7)

The export price of plywood is affected by the previous year's export price of plywood and plywood export of the current year. When the former increases 1%, the export price of plywood increases 0.89% (AG2). When the latter increases 1%, the export price increases 0.15% (AG3).

THE SOFTWOOD MARKET

Historically, softwood species, especially the yellow cedars (*Chamaecyparis* spp.), have been the favorite species in local wood utilization. Over the years, the softwood forests have, therefore, been heavily harvested. In a move to stress the importance of nontimber functions of the forest, a new forest policy was implemented in 1977 to reduce the level of timber harvest (both softwood and hardwood). Because of the popularity of the yellow cedars in the Japanese market, softwood logs have long been exported. Even today some softwood logs are being exported, even though the softwood forests are no longer as abundant. In the analysis below, the softwood market consists of (1) production of softwood logs, (2) consumption of softwood logs, (3) softwood log prices, and (4) export of softwood logs.

Production of Softwood Logs (Equation 9)

Softwood log production is a function of the softwood log price, black market interest rate, and logging wage rate. In addition, the energy crisis of 1973 and the new forest policy since 1977 have affected softwood production. As shown in Table 2, when softwood log price increases 1%, softwood log production increases 0.16% (BA2). Therefore, softwood timber supply is rather price inelastic. When the black market interest rate and logging wage rate increase 1%, softwood log production decreases 0.40% and 0.28% (BA3 and BA4), respectively. That is to say, softwood log production does not vary much as a result of changes in the cost of these two factors. Since the 1973 energy crisis, the softwood log production has been reduced by about 7% every year. The 1977 forest policy has meant a 15% reduction of softwood production since its implementation.

Softwood Log Consumption (Equation 10)

Softwood consumption is a function of housing starts in Taiwan, softwood log price, imported hardwood log price, and a long-term trend. Results in Table 2 show the long-term trend to be the only significant variable every year the consumption of softwood decreases about 7%. The effect of housing starts on softwood log consumption is statistically insignificant: a 1% increase in the former causes only a 0.22% (BB2) increase in the latter. The price responses of softwood consumption to changes in both softwood log price and imported hardwood log price (BB3 and BB4) are both of the wrong sign and statistically insignificant. Thus it is safe to say that neither price change has much effect on softwood log consumption.

Softwood Log Prices (Equation 11)

The softwood log price depends mainly on the previous year's softwood log price, the price of imported hardwood logs, and the softwood log consumption. A 1% increase in the latter would

Table 2. Parameters associated with equations of softwood markets.

Equation	Parameter	Estimate	Approximate Standard Error	t Value
9	BA1	15.96	0.74	21.48**
	BA2	0.16	0.09	1.81
	BA3	−0.40	0.20	−2.03
	BA4	−0.28	0.05	−5.45**
	BA5	−0.06	0.06	−0.94
	BA6	−0.16	0.06	−2.87*
10	BB1	12.64	0.97	13.04**
	BB2	0.22	0.16	1.37
	BB3	0.04	0.21	0.21
	BB4	−0.16	0.16	−0.97
	BB5	−0.07	0.02	−2.69*
11	BC1	−1.88	2.52	−0.75
	BC2	0.63	0.23	2.69*
	BC3	0.30	0.26	1.15
	BC4	0.18	0.19	0.96
12	BD1	−5.67	10.83	−0.52
	BD2	−1.32	0.44	−3.00**
	BD3	1.55	0.92	1.69
	BD4	−1.45	1.53	−0.95
	BD5	-0.34	0.64	−0.53

* 5% significance.
** 1% significance.

lead to a 0.63% (BC2) increase in the former. The imported hardwood log price and softwood log consumption exert only a small influence on the softwood log price. A 1% increase in the imported hardwood log price increases the softwood log price by 0.30% (BC3), and a 1% increase in the softwood log consumption causes a 0.18% (BC4) increase in the softwood log price. Both of these effects are not statistically significant.

Export of Softwood Logs (Equation 12)

Softwood logs exported have mainly been yellow cedars. Traditionally, Japan has been the export market. Therefore, softwood log export depends on export price of softwood logs, Japan's housing starts, and the Japanese yen/Taiwan dollar exchange rate. When export price of softwood logs increases 1%, softwood export decreases 1.32% (BD2). It is, therefore, fair to say that Japanese import demand for Taiwan yellow cedar is quite price elastic. Other factors, while not statistically significant, are nevertheless important. For example, a 1% increase in Japanese housing starts leads to a 1.55% (BD3) increase in softwood log exports, and a 1% appreciation of the Taiwan dollar against the Japanese yen would lead to a 1.45% (BD4) drop in the export of softwood logs. Suffice it to say, the export of softwood logs is dictated entirely by the Japanese market. The supply side is in no position to bargain. During 1983 and 1984, some unusual activities were going on in terms of indirect exports of softwood logs from Taiwan to mainland China. A dummy variable was introduced to isolate the effect of such an event. Surprisingly, the coefficient turns out to be negative and not statistically significant. Maybe indirect export activities are just not that significant after all.

THE HARDWOOD MARKET

The hardwood market consists of (1) production of hardwood logs, (2) price of hardwood logs, (3) import of hardwood logs, and (4) import of hardwood lumber. The entire hardwood market is dominated by the import of hardwood logs and lumber. The results of the parameter estimates for the hardwood markets are given in Table 3.

Table 3. Parameters associated with equations of hardwood markets.

Equation	Parameter	Estimate	Approximate Standard Error	t Value
14	CA1	19.65	3.60	5.46**
	CA2	−0.05	0.58	−0.08
	CA3	−1.99	0.91	−2.18*
	CA4	−0.05	0.19	−0.25
	CA5	0.16	0.33	0.48
	CA6	−0.29	0.21	−1.39
15	CB1	0.13	0.08	1.58
	CB2	0.41	0.14	2.97**
	CB3	0.47	0.14	3.37**
16	CC1	0.58	4.13	0.14
	CC2	0.98	0.27	3.60**
	CC3	0.21	0.27	0.79
	CC4	−0.11	0.18	−0.62
	CC5	0.23	0.43	0.53
	CC6	−0.32	0.22	−1.50
	CC7	0.16	0.20	0.82
17	CD1	−10.53	1.32	−7.95**
	CD2	4.47	0.39	11.56**
	CD3	−0.41	0.12	−3.33**
	CD4	−2.51	0.49	−5.12**
	CD5	0.90	0.25	3.65**
	CD6	0.45	0.29	1.55
	CD7	−0.66	0.26	−2.49**

* 5% significance.
** 1% significance.

Production of Hardwood Logs (Equation 14)

The production of hardwood logs is affected by hardwood price, black market interest rate, and logging wage rate. In addition, the 1973 energy crisis and 1977 new forest policy exert certain influences on the production of hardwood logs. The effect of hardwood price on hardwood log production is slightly negative and thus essentially none. The black market interest rate exerts considerable influence on the production of hardwood: a 1% increase in the former would lead to a 2% (CA3) drop in the latter. The effect of the logging wage rate on the other hand is much smaller: a 1% increase in logging wage rate would lead to only a 0.05% (CA4) drop in the production of hardwood logs. Since the energy crisis of 1973, the production of hardwood logs is about 17% higher than without the energy crisis. There are two possible explanations for this result. First, most hardwood forests are at lower elevations and therefore closer to the market, with lower transportation costs. As a result of the higher energy cost since the energy crisis, hardwood forests became more attractive and thus more is harvested. Second, the coefficient is not statistically significant; therefore, the effect of the energy crisis is really not all that significant. The new forest policy since

1977 brought about a 25% reduction of hardwood log production.

Price of Hardwood Logs (Equation 15)

The price of hardwood logs is affected by the previous year's price of hardwood logs and the import price of hardwood logs. A 1% increase in the previous year's price would cause a 0.41% (CB2) increase in the current year. In comparison, a 1% increase in the import price of hardwood would lead to a 0.47% (CB3) increase in the price of hardwood logs. Obviously the significance of imported hardwood logs cannot be overlooked.

Import of Hardwood Logs (Equation 16)

The import of hardwood logs dominates the entire softwood and hardwood market. In any particular year, imported logs (almost entirely hardwood) consist of around 85 to 90% of the entire market. The import of hardwood depends on the production of plywood, the value of furniture production, the real black market interest rate, wage rate in the wood products industry, price of industrial fuel oil, and the import price of hardwood logs. The relationship between these variables and the import of hardwood logs is as follows:

A 1% change in	results in a change of hardwood log imports of
Plywood production (PROPLY)	0.98% (CC2)
Value of furniture production (PROFUR)	0.21% (CC3)
Real black market interest rate (RBMR)	−0.11% (CC4)
Wage rate in wood products industry (WRWPI)	0.23% (CC5)
Price of industrial fuel oil (PRFUEL)	−0.32% (CC6)
Import price of hardwood logs (PRMHLG)	0.16% (CC7)

Of all the variables, only the coefficient for production of plywood is highly significant. The real black market interest rate and the price of industrial fuel oil are not statistically significant. The positive coefficient of the import price of hardwood logs seems to vindicate the adage "Use foreign resources to earn foreign exchange," because the higher import price of hardwood logs usually also signals a better export climate.

Import of Hardwood Lumber (Equation 17)

The import of hardwood lumber depends on the value of wood furniture production, real black market interest rate, wage rate in the wood products industry, price of fuel oil, import price of hardwood logs, and import price of hardwood lumber. Amazingly, for every 1% increase in the value of wood furniture production, the import of hardwood lumber would increase 4.47% (CD2). Clearly, as the furniture industry in Taiwan continues to grow, more and more lumber, rather than logs, will be imported. The negative coefficient of the real black market interest rate probably reflects the potent effect of the interest rate in bringing down the value of furniture production, which in turn reduced the demand for the import of hardwood lumber. The −2.51 (CD4) coefficient for the wage rate in the wood products industry may be attributed to the quickness that the labor cost responds to business conditions. Finally, it is important to observe the substitutional relationship between hardwood logs and lumber. It is also worth pointing out that the import demand for hardwood lumber is quite price inelastic.

MODEL SIMULATION

Given this simultaneous equations model, it is interesting and important to investigate how well the model will perform in simulating the actual events. Table 4 presents the results of the analysis of the simulation efforts. While no single number shows the whole picture, together they do present

Table 4. Results of simulation analysis.

Variable	N	RMS Error	RMS Error (%)	MSE BIAS	Decomposition Reg.	Dist.	R^2
FURNITURE AND PLYWOOD INDUSTRIES							
EXFUR	24	41.07	431.88	.005	.000	.995	.9333
PROFUR	24	737.76	95.00	.115	.005	.880	.9563
PROPLY	24	238379	36.57	.197	.041	.762	.7405
CONPLY	24	52157.95	102.89	.033	.076	.891	.9143
PRIDPLY	24	10.37	22.42	.198	.113	.688	.8340
EXPLY	24	1652.01	32.08	.072	.011	.918	.7588
PRXPLY	24	82.97	44.30	.461	.137	.402	.6042
SOFTWOOD AND HARDWOOD MARKETS							
EXSLG	23	13390.41	45.52	.003	.058	.938	.6667
PRHLG	23	.40	43.70	.090	.143	.767	.8420
IMHLG	23	334366	28.01	.034	.061	.905	.9718
IMHLR	19	5952.69	125.72	.018	.009	.974	.9980
CONSLG	23	53151.24	10.37	.010	.029	.961	.8293
PRSLG	23	1.46	52.99	.114	.317	.569	.7231
PROHLG	23	55062.56	17.58	.050	.040	.910	.6033
PROSLG	23	83685.44	13.40	.292	.386	.322	.7262
CONHLG	23	323296	15.08	.019	.037	.944	.9729

a fairly complete one. First, one should not be alarmed by the results involving the plywood imports (IMPLY) and softwood log imports (IMSLG). These two variables serve in the model as the variables that balance the supply and demand relation. The other identity variable, domestic consumption of hardwood logs (CONHLG), performed very well because of the success of the IMHLG equation in describing the import of hardwood logs. Based on R^2, the equations of the two wood product industries, the production of plywood (PROPLY) and the export of plywood (EXPLY), would require further improvement. In addition, the two price equations for price index of plywood (PRIDPLY) and export price of plywood (PRXPLY) are not satisfactory. Clearly, alternative forms of price formation need to be explored. The plywood production and export situation may be an entirely different story. With rapid changes in the hardwood plywood industry internationally, the plywood industry in Taiwan is rapidly undergoing qualitative transformation. For example, Taiwan began charging antidumping, countervailing duty on Indonesian plywood imports recently, essentially protecting the domestic plywood industry. Thus, what was once a giant industry may soon become a minor one in Taiwan. It may be a few more years before more definite models for these two equations will emerge.

For the softwood and hardwood markets, by and large the entire model performed rather well. However, it would appear that the price and production of softwood (PRSLG and PROSLG) can be improved upon in the model.

CONCLUSION

In this paper, a simultaneous equations model for the furniture and plywood industries as well as the softwood and hardwood markets is presented. From a technical point of view, it is clear that a decent model can be developed with supply and demand equations derived from well-behaved and well-understood production functions. From a practical point of view, the decline of the plywood industry and the rise of the wood furniture industry merit some observation. As the plywood industry continues to decline, the amount of hardwood peeler logs imported will also decline. As such, the amount of lumber produced from peeler cores and rejected peeler logs will also be reduced. Such a development is already having an effect on imports of hardwood lumber. The rapid growth of the wood furniture industry simply stokes the fire even further, as evidenced by the 4.47% increase in the import of hardwood lumber for every 1% increase in the value of wood furniture production. Export suppliers in the United States would be well advised to carefully follow the development of the wood furniture industry in Taiwan and be prepared to fight for their share of the market.

ACKNOWLEDGMENT

The research reported in this paper was carried out while the author was a visiting specialist at the Taiwan Forestry Research Institute. The research was supported in part by a research grant from the Council of Agriculture of the Republic of China.

REFERENCES

Chang, S. J., and I-an Jen. 1986. An econometric analysis of the timber supply and demand situation in Taiwan. Project report submitted to the Council of Agriculture. (In Chinese).

SAS Institute Inc. 1984. SAS/ETS User's guide. SAS Institute Inc., Cary, North Carolina.

Varian, H. R. 1978. Microeconomics analysis. W. W. Norton and Co., New York.

APPENDIX 1. Various statistics of the Taiwan economy from 1961 to 1984.

Year	Per Capita GNP (U.S.$)	Internal Log Production (m³)	Log Imports (m³)	Plywood Production (MSF, 1/8")
1961	151.15	897,978	168,042	219,701
1962	161.63	904,722	276,716	296,441
1963	177.55	879,026	446,506	515,530
1964	201.85	1,069,582	565,191	770,510
1965	216.38	1,116,915	625,141	847,354
1966	235.90	1,007,010	691,917	1,002,982
1967	265.93	1,060,462	741,670	1,015,708
1968	302.43	1,118,215	1,093,474	1,476,587
1969	342.98	1,063,563	1,183,025	2,064,100
1970	386.68	1,109,943	1,489,333	2,576,097
1971	440.85	1,217,291	2,205,287	2,978,957
1972	519.03	1,136,145	3,568,130	4,487,889
1973	696.81	1,099,186	3,830,478	4,550,619
1974	913.94	982,971	3,523,227	3,733,324
1975	957.05	854,731	3,622,917	3,533,741
1976	1,123.32	820,694	3,917,386	4,158,989
1977	1,289.54	689,435	5,428,096	4,312,554
1978	1,602.70	674,107	6,642,944	5,448,607
1979	1,896.39	653,529	6,290,754	4,918,875
1980	2,314.07	582,138	4,968,330	4,407,260
1981	2,561.97	529,684	5,205,679	4,644,574
1982	2,548.92	494,937	4,472,766	4,203,848
1983	2,732.99	616,070	4,790,766	4,237,750
1984	3,081.35	562,637	4,106,375	3,390,200

APPENDIX 2. List of variables for the solid-wood products model for Taiwan and the equations used in the model.

Variable	Description

ENDOGENOUS VARIABLES

EXFUR	Export of furniture ($1,000)
PROFUR	Production of furniture ($1,000)
PROPLY	Production of plywood (m^3)
CONPLY	Consumption of plywood (m^3)
PRIDPLY	Price index of plywood (1981 = 100)
EXPLY	Export of plywood (m^3)
PRXPLY	Price of exported plywood
IMPLY	Import of plywood
PROSLG	Production of softwood logs (m^3)
CONSLG	Domestic consumption of softwood logs (m^3)
PRSLG	Price of softwood logs
PRXSLG	Price of exported softwood logs
EXSLG	Export of softwood logs (m^3)
IMSLG	Import of softwood logs (m^3)
PROHLG	Production of hardwood logs (m^3)
PRHLG	Price of hardwood logs (m^3)
IMHLG	Import of hardwood logs (m^3)
IMHLR	Import of hardwood lumber (m^3)
CONHLG	Domestic consumption of hardwood logs (m^3)

EXOGENOUS VARIABLES AND LAGGED ENDOGENOUS VARIABLES

RGNPUSA	Real GNP of U.S. in 1977 dollars (million U.S.$)
RBMR	Real black market interest rate (%)
WRWPI	Wage rate in wood products industry
PRELEC	Price of electricity for industrial users
PRFUEL	Price of industrial fuel oil
HSROC	Housing starts in Taiwan
PRIDPLY(−1)	Price index of plywood lagged one year
PRXPLY (−1)	Price of exported plywood lagged one year
D4	A dummy variable for before and after energy crisis. Before 1973 (1973 not included) D4 = 0. After 1973, D4 = 1.
D5	A dummy variable for the new forest policy. Before 1977 (1977 not included) D5 = 0. After 1977, D5 = 1.
D6	A dummy variable for the years of indirect export of softwood logs to mainland China. D6 = 1 for 1983 and 1984. All other years, D6 = 0.
BMR	Black market nominal interest rate (%)
WRLOG	Wage rate of the logging industry
PRSLG(−1)	Price of softwood logs lagged one year
HSJAP	Housing starts in Japan
ERJAP	Exchange rate for Japanese yen per dollar new Taiwan currency (NTC)

APPENDIX 2. Continued.

Equation Number	Functional Form

FURNITURE INDUSTRY

1. EXFUR

$EXP(AA1 + AA2/RGNPUSA)$

2. PROFUR

$EXP(AB1) (EXFUR/35.75)^{AB2} RBMR^{AB3} WRWPI^{AB4}*$
$PRELEC^{AB5} PRMHLG^{AB6}$

PLYWOOD INDUSTRY

3. PROPLY

$EXP(AC1 + AC7/(YR-1952)) PRIDPLY^{AC2} RBMR^{AC3}*$
$WRWPI^{AC4} PRFUEL^{AC5} PRMHLG^{AC6}$

4. CONPLY

$EXP(AD1 + AD4/(YR-1952)) HSROC^{AD2} PRIDPLY^{AD3}$

5. PRIDPLY

$EXP(AE1) PRIDPLY(-1)^{AE2} CONPLY^{AE3} EXPLY^{AE4}$

6. EXPLY

$EXP(AF1) PRXPLY^{AF2} PROPLY^{AF3}$

7. PRXPLY

$EXP(AG1) PRXPLY(-1)^{AG2} EXPLY^{AG3}$

8. IMPLY

$CONPLY + EXPLY - PROPLY$

SOFTWOOD MARKET

9. PROSLG

$EXP(BA1 + BA5*D4 + BA6*D5) PRSLG^{BA2} BMR^{BA3}$
$WRLOG^{BA4}$

10. CONSLG

$EXP(BB1 + BB5*(YR-1952)) HSROC^{BB2} PRSLG^{BB3}$
$PRMHLG^{BB4}$

11. PRSLG

$EXP(BC1) PRSLG(-1)^{BC2} PRMHLG^{BC3} CONSLG^{BC4}$

12. EXSLG

$EXP(BD1 + BD5*D5) PRXSLG^{BD2} HSJAP^{BD3} ERJAP^{BD4}$

13. IMSLG

$CONSLG + EXSLG - PROSLG$

HARDWOOD MARKET

14. PROHLG

$EXP(CA1 + CA5*D4 + CA6*D5) PRHLG^{CA2} BMR^{CA3}$
$WRLOG^{CA4}$

15. PRHLG

$EXP(CB1) PRHLG(-1)^{CB2} PRMHLG^{CB3}$

16. IMHLG

$EXP(CC1) PROPLY^{CC2} PROFUR^{CC3} RBMR^{CC4} WRWPI^{CC5}*$
$PRFUEL^{CC6} PRMHLG^{CC7}$

17. IMHLR

$EXP(CD1) PROFUR^{CD2} RBMR^{CD3} WRWPI^{CD4} PRFUEL^{CD5}*$
$PRMHLG^{CD6} PRMHLR^{CD7}$

18. CONHLG

$PROHLG + IMHLG$

Let K, L, E, and M represent the capital, labor, energy, and raw material; then the output quantity Q can be expressed as

$$Q = H * K^a * L^b * E^c * M^d \tag{1}$$

where H, a, b, c, and d are all parameters for generalized Cobb-Douglas production function. From this simple production function, one can then derive the corresponding cost function (Varian 1978)

$$C(Q, PK, PL, PE, PM) = Z * Q^{1/s} * (PK)^{a/s} * (PL)^{b/s} * (PE)^{c/s} * (PM)^{d/s} \tag{2}$$

where Z is another parameter and PK, PL, PE, and PM are the costs of capital, labor, energy, and raw material respectively. Note that $s = a + b + c + d$ is an indicator of the economy of scale. When $s > 1$, the industry under investigation possesses increasing return to scale; when $s = 1$, it represents constant return to scale; and when $s < 1$, it represents decreasing return to scale. Applying Shepard's Lemma, the partial derivative of the cost function with respect to the cost of raw material represents the factor input demand function when the output is Q:

$$M = \frac{\partial C}{\partial(PM)} = (\frac{Z*d}{s}) * Q^{1/s} * (PK)^{a/s} * (PL)^{b/s} * (PE)^{c/s} * (PM)^{-(a+b+c)/s} \tag{3}$$

When the cost function mentioned above is inserted into the profit function π, then

$$\pi = P * Q - C(Q, PK, PL, PE, PM) \tag{4}$$

where P is the price of the output and everything else is as mentioned above. Taking partial derivative of the profit function with respect to Q and setting it equal to zero gives

$$\frac{\partial \pi}{\partial Q} = P - (\frac{Z}{s}) * (PK)^{a/s} * (PL)^{b/s} * (PE)^{c/s} * (PM)^{d/s} * Q^{(1-s)/s} = 0 \tag{5}$$

and consequently Q can be solved for

$$Q = (\frac{S}{Z})^{st} * P^{st} * (PK)^{-at} * (PL)^{-bt} * (PE)^{-ct} * (PM)^{-dt} \tag{6}$$

with $t = 1/(1-s)$.

Chapter 7

The Market for Solid-Wood Products in Korea: Past and Future Opportunities

KEITH A. BLATNER, GERARD F. SCHREUDER, ROBERT L. GOVETT, and YEO YOUN

The Republic of Korea is now the fourth largest export market for United States wood products, ranking behind Japan, Canada, and China. Wood shipments from the United States to Korea totaled $137.3 million ($U.S., f.a.s., port of export) in 1986. This was the largest total dollar value of exports to Korea in this decade and represented an increase of 16% over 1985 levels. This figure may appear small compared with the total annual volume of U.S. wood shipments to Japan of $1,286.6 billion; but it represents a third of the value of shipments to Canada and 75% of the value of shipments to China during the same year. This market has exhibited a steady pattern of growth over the past six years, with imports of U.S. wood products rising from $59.4 million in 1981 to $137.3 million in 1986, an increase of 14.98% compounded annually (USDA 1987). These changes are particularly dramatic in light of the large declines in the hardwood plywood and sawmill sectors in Korea over the same period.

A National Forest Products Association "white paper" (NFPA 1985) found Korea to have significant potential for increased consumption of U.S. wood exports. Expectations for future growth in this market are based on a rapidly expanding economy, an expanding housing market, the changing Southeast Asian raw material supply situation, and an indicated willingness on the part of the Republic of Korea to further open its markets.

Although recent research has increased our knowledge of this market, it is important to recognize the many factors to be considered when evaluating it. The objective of this paper is to provide readers with a greater understanding of the Korean market for softwood and hardwood products, concentrating on future demand for logs and lumber. The authors have purposely avoided discussing the hardwood plywood industry, because of its rapidly changing nature within the country over the past few years. Two estimates of potential market growth through the year 2000 and the potential market roles of alternative suppliers in servicing the Korean market in the near future are also considered.

The data for this analysis were obtained courtesy of the Forest Research Institute (FRI), Seoul, South Korea, and as a result of a recently completed research project designed to access current forest resource management within Korea and project future demand for various forest products (Schreuder 1986). Baseline data on forest products consumption by industry and species were obtained from *An Illustration of Forest Products Utilization* (FRI 1982).

DOMESTIC PRODUCTION AND IMPORTS

Only about one-fifth of the 7.1 million cubic meters (m^3) of the wood consumed in Korea during 1981 (FRI 1982), the latest year for which a detailed survey of internal consumption was conducted, came from domestic sources. Because of the limitations of its domestic timber supply,

Korea imported over 80% of the total volume of wood products consumed. Hardwoods accounted for 71.5% of the total imports in 1981, and softwoods for 28.5%.

As a direct result of deforestation during the Korean War, only about 3% of Korea's 177 million m^3 of growing stock exceeds 60 years of age. About 70% of all growing stock is under 30 years of age. These age distribution problems are further complicated by species, utilization, and ownership problems. Currently, 47.7% of the total growing stock is made up of Korean red pine (*Pinus densiflora*), which is generally of poor form. Approximately 27% of the remaining growing stock is made up of various oaks (primarily *Quercus serrata, Q. acutissima*, and *Q. grosseserrata*), which are also generally small and of poor form. Other, potentially more useful species such as Korean white pine (*Pinus koraiensis*), birch (*Betula* spp.), basswood (*Tilia amurensis*), ash (*Fraxinus rhynchophylla*), and paulownia (*Paulownia coreana*) are in limited supply, accounting for only 2% of Korea's harvest during recent years. About 85% of the domestic supply is provided by small private forest holdings, with the remainder coming from national forest lands.

Malaysia was the largest supplier of solid-wood products during 1981, accounting for over 61.6% of all imports. The United States was the second largest supplier, with 16.3% of the market, followed by Indonesia (8.7%), Papua New Guinea (PNG) (4.7%), Philippines (2.8%), Chile (4.3%), and others (1.6%). Imports of tropical hardwoods included apitong and keruing (*Dipterocarpus* spp.), jelutong (*Dyrea costulata*), kapur (*Dryobalonops* spp.), lauan (*Shorea* spp.), sepetir (*Sindora wallichi*), and teak (*Tectona grandis*). Korean imports from the United States included western hemlock (*Tsuga heterophylla*), Douglas-fir (*Pseudotsuga menziesii*), walnut (*Juglans* spp.), white fir (*Abies concolor*), spruce (*Picea* spp.), oak (*Quercus* spp.), and maple (*Acer* spp.). Imports from Chile were primarily radiata pine (*Pinus radiata*).

RESIDENTIAL AND NONRESIDENTIAL CONSUMPTION

During 1981, the building industry used 1.75 million m^3 of wood, primarily in such interior applications as door frames, moldings, and other joinery products. Approximately two-thirds of the volume of wood used in building construction was for residential construction, about one-fifth was consumed in commercial construction, and the remainder was split in use among industrial, public, and other types of construction (Table 1).

Table 1. Korean wood consumption by building type in 1981.

Building Type	Building Area (1,000 m^3)	Total Consumption (1,000 m^3)
Private	10,308	1,110
Commercial	4,959	331
Industrial	2,647	133
Public	1,856	109
Other	1,076	67
Total	20,846	1,750

Less than 4% of the wood used in Korean building construction is of domestic origin. Lauan overshadows all other species combined in popular use, accounting for 67% of all the wood used in building construction on a volume basis. Softwoods accounted for almost 30% of the volume of wood used in building construction. Although specific species/product breakdowns were not available, use of Douglas-fir for joinery products is reported to have gained an increasing share of the market in recent years.

INDUSTRIAL CONSUMPTION

Industrial consumption of solid-wood products in Korea is concentrated in such diverse applications as subway construction, packing cases, the manufacture of musical instruments and furniture, and various agricultural applications, among others. Of these industries, the furniture and musical instrument industries are the most important from a value-added standpoint; however, subway construction ranked number one based on the total volume of wood products consumed (Table 2).

Table 2. Korean consumption of wood products by end use in industrial applications in 1981.

End Use	Consumption (100 m^3)
Subway construction	500.0
Furniture	104.0
Pallets and shipping container materials	99.5
Packing cases	94.0
Musical instruments	93.2
Seaweed timbers (cultivation beds)	49.8
Mushroom timbers (cultivation beds)	43.0
Vinyl house construction (greenhouses)	21.0
Ginseng tea growing yards	19.5
Tool handles	14.7

Korea's musical instrument industry has grown considerably in recent years, with piano manufacturing nearly doubling since 1983. Korea is expected to surpass Japan in piano production in the near future (FRI, pers. comm., 1986). The United States is one of Korea's more important markets for musical instruments, buying an estimated 30,000 Korean pianos during 1986. Korea's piano manufacturers are producing instruments for the mass market as well as very high quality instruments. Korea is also an important manufacturer of guitars.

Korea's musical instrument manufacturers consumed in excess of 90,000 m^3 of solid wood during 1981, of which almost 85% was for pianos. This surprisingly large volume has undoubtedly increased dramatically along with a doubling in piano manufacturing capacity since that time. The primary species used by Korean piano manufacturers were lauan, spruce (*Picea* spp.), and maple, in order of importance on a volume basis, as well as oak, walnut, and birch. Most of the remaining wood used in the manufacture of musical instruments is for guitars, with kalopanax (*Kalopanax pictum*), maple, lauan, and spruce being the most important species in terms of volume.

More than 98% of the 103,900 m^3 of wood used in the manufacture of furniture during 1981 was imported. Historically, tropical hardwoods have accounted for more than half the volume of solid wood consumed, with lauan accounting for approximately 40% of the total volume of solid wood consumed in furniture production. Other tropical species used include sepetir, apitong, teak, agathis (*Agathis alba*), bintangor (*Calophyllum* spp.), and jelutong. Walnut and oak from the United States accounted for slightly more than 10% of the total volume used, with maple (*Acer* spp.), birch, beech (*Fagus* spp.), elm (*Ulmus* spp.), ash, and cherry (*Prunus* spp.) also being consumed in lesser quantities. Western hemlock from the United States accounted for about 5.5% of the volume of solid wood used by Korea's furniture manufacturers.

About 500,000 m^3 of wood were used in subway construction in Korea in 1981. Of this volume, 73% was keruing, a tropical hardwood imported from Southeast Asia. Softwoods and domestic larch from the United States accounted for about 20% of the remainder. Subway construction is not only the largest industrial market for wood products on a volume basis but also represents the single largest industrial use of U.S. softwoods in Korea.

Korea annually consumes about 94,000 m^3 of wood in the manufacture of cable drums and packing cases. Imports of radiata pine and U.S. softwoods accounted for 46.8% and 32.8% of this market, respectively. Domestically produced species, particularly Korean red pine, accounted for the rest. Relative market shares varied by type of packing case. Softwoods and radiata pine from the United States were the dominant species in the production of cable drums; U.S. softwoods accounted for 100% of Korean shipping cases. Conversely, Korean red pine was the principal species used in the manufacture of ginseng packing cases, with U.S. softwoods providing the rest.

An additional 100,000 m^3 were consumed in the manufacture of pallets and shipping containers, with kapur and keruing lumber used almost exclusively in the manufacture of truck flooring. Both lumber and plywood were used in the production of shipping container flooring. About 20,000 m^3 of lumber were used in truck flooring, while about 50,000 m^3 of plywood and 30,000 m^3 cubic of lumber were consumed in container manufacturing.

Korea annually consumes about 40,000 m^3 of wood for use as timbers for mushroom cultivation beds. Approximately 90% is derived from Korean species. Mushroom timbers range from 6 to 21 centimeters in diameter and about 1.0 to 1.5 meters in length. Duration of use is in direct proportion to size, with smaller timbers lasting three to four years and larger timbers six to eight years.

Korean farms annually consume about 21,000 m^3 of lumber in the construction of "vinyl houses" (greenhouses covered with plastic instead of glass) and 42,000 m^3 of lumber in ginseng tea growing yards. Lumber used in the construction of vinyl houses is typically of square dimensions of 3.0 by 3.0 or 4.5 by 4.5 centimeters and 1.8, 2.55, 2.7, and 3.6 meters in length. Keruing is the most commonly used species in vinyl house construction, while chestnut (*Castanea* spp.), keruing, and locust (*Robinia* spp.) are used in ginseng yards.

Approximately 50,000 m^3 of wood are consumed annually as seaweed timbers in Korea's 25,000 hectares of seaweed cultivation beds. Log diameters of 3 to 8 centimeters are preferred, with lengths of 4 to 15 meters commonly used. Domestic pines, larch, and oak are used as seaweed timbers, with significant premium paid for oak in lengths greater than 10 meters.

About 15,000 m^3 of wood are consumed annually in the production of tool handles, with slightly over half the volume supplied by domestic producers. Domestic species used include oak, maple, ash, and other hardwoods. More than 6,000 m^3 of tropical hardwoods are also used in the manufacture of tool handles. No evidence of the use of temperate hardwoods for tool handles could be found.

MARKET FACTORS

Given Korea's growing economy and increasing standard of living, it is only logical to expect demand for both hardwoods and softwoods to expand. However, as with any market, there are numerous factors to be considered.

Future Korean demand for U.S. products will be tempered by a desire to establish new long-term sources of supply; U.S. producers should expect stiff competition from other suppliers. Exporters must realize that even though the Korean market is in a period of transition and holds numerous opportunities, Koreans are shrewd businessmen mainly interested in obtaining good value for each expenditure.

Recent changes in the availability of tropical hardwood logs will continue to cause serious log supply problems. As Korea's forest products industry continues its restructuring over the next decade, it is projected that demand for sawn tropical hardwoods will increase (Govett et al. 1987). Further, since Korean industry is already accustomed to these species, U.S. producers will have to convince Korean manufacturers and consumers that U.S. wood products represent a better value, at least in applications where tropical hardwoods have historically dominated.

Competition from radiata pine producers, particularly Chile, will also become a more important factor as existing plantations become merchantable. Korea has established a strong market link with Chile over the past few years, which will undoubtedly continue to grow. Korea was Chile's second largest forest products market during 1985, with imports of $29.4 million ($U.S., f.o.b., Chile) (Anon. 1986). Although this figure includes pulp and paper and a variety of lesser forest products, it does show the importance of Chile as a Korean supplier. New Zealand is often thought of as the largest supplier of radiata pine logs in the Pacific Rim; however, Chile's exports of radiata pine logs to Korea have exceeded New Zealand's by a factor of 3.5 in every year since 1978. As an example, Chile's exports to Korea totaled 461,562 m³ compared with New Zealand's exports of 124,974 m³ during 1984 (Blatner et al. 1987). While it is recognized that radiata is generally of lower quality than many other species, particularly with respect to strength, recent research has shown that radiata pine can be used effectively in the manufacture of furniture and in other non-structural applications. This species is also easily treated for use in applications where durability is a key factor.

During 1984, the Korean government reduced the ad valorem tariff on softwood sawlogs from 10% to 5%. Tariffs on sawn and other processed material were reduced from 30-40% to 20%. During January 1987, tariffs on softwood sawn products were again reduced from 20% to 15%. Even so, a 15% ad valorem tariff represents a substantial barrier to increased trade (NFPA 1987).

The negative aspects associated with the Korean tariff on imported lumber may be avoided, or at least minimized, if U.S. exporters are able to target specific industries where the final product is reexported. The essence of this rebate procedure is outlined in the following quotation:

> Korean importers can receive a refund of ad valorem import tariffs charged, in proportion to the volume of imported wood that is reexported, based on adjustment by the expected recovery ratio of finished product to the particular form of raw material used. For example, if a Korean furniture manufacturer imports 100 m³ of hemlock lumber, for which a 20% tariff is charged against the value, and assuming a 70% normal recovery rate may be expected in the production of finished furniture stock from lumber of such a grade, the furniture manufacturer could apply for a rebate of the tariff charged against the full 100 m³ of lumber as long as 70 m³ of the hemlock were components in exported furniture. If 35 m³ of the imported hemlock were present in the exported furniture, the importer could apply for a rebate of half the tariff paid. (Blatner et al. 1987)

Obviously, this aspect of the Korean import tariff system does not nullify the effect of the tariff for products destined for the Korean market; however, it is an important factor to be considered when evaluating alternative market opportunities both inside and outside Korea.

FUTURE MARKET POTENTIAL

There are two important aspects of any market assessment—projected market size and market opportunities. For a better understanding of the potential size of this market, two projections of growth in the demand for solid-wood products have been constructed based on projected growth in the Korean economy as measured by per capita gross national product, population, and expected U.S. market share. The first scenario assumes that U.S. producers are successful in increasing their current market share from approximately 16% to 40% by the year 2000. The second scenario addresses the future potential of this market if the United States is unable to expand its share of this market, because of competition from other sources of supply, but is able to maintain its current market share. Potential market opportunities are addressed in subsequent sections of the paper.

A series of simple models were considered in projecting future sawnwood consumption in Korea. Based on these efforts, the authors estimate that total sawnwood consumption in Korea will reach 4.7 million m³ by the year 2000 (Table 3). Although this total represents an increase of more than 100% over the 1981 level, it is substantially less than the amounts for the peak years of 1978 and 1979 and represents a moderate increase over the historically high levels of 1983 and 1984

Table 3. Estimated sawnwood consumption and related indicators for Korea, 1987-2000.

Year	Projected Sawnwood Consumption* (1,000 m³)	Projected per Capita GNP** (1,000 won)	Projected Total Population** (million)
1987	2,271.3	2,091.79	42.4
1988	2,398.8	2,283.33	43.0
1989	2,532.7	2,492.84	43.6
1990	2,668.8	2,735.49	44.1
1991	2,813.7	3,002.35	44.6
1992	2,975.1	3,208.36	45.2
1993	3,139.7	3,610.15	45.7
1994	3,322.5	3,955.33	46.3
1995	3,517.4	4,334.26	46.9
1996	3,719.6	4,760.27	47.4
1997	3,931.2	5,239.71	47.8
1998	4,165.7	5,755.86	48.3
1999	4,490.8	6,523.55	48.8
2000	4,701.1	6,945.20	49.4

*Estimated sawnwood consumption is based on the following estimated relationship: $Y = (-2616.12) + (97.51)X_1 + (0.36)X_2$, where Y = sawnwood consumption, X_1 = population, X_2 = per capita GNP. ($R^2 = 0.87$, F = 67.53, n = 14).
**Source: Korea Development Institute.

(Figure 1). It is possible to generate much higher estimates of future consumption; but this relatively conservative estimate appears reasonable in light of recent government efforts to limit or at least exert some control over the total volume of wood imports.

Based on these projections, imports of U.S. sawnwood products and logs after manufacture would total 1.88 million m³ by the year 2000 under the first scenario, and 752,000 m³ under the second.

IMMEDIATE AND LONGER-TERM MARKET OPPORTUNITIES

The greatest short-run opportunity for West Coast producers appears to lie in increased softwood log exports. This observation reflects the large potential capacity of the existing sawmill industry and Korea's long history of importing logs instead of lumber. Similarly, the best short-run opportunities for hardwood producers appear to be in supplying Korea's healthy and growing furniture and musical instrument industries. Hardwood exporters should concentrate initially on the export of higher value species such as walnut, oak, and maple, which Korean manufacturers are already familiar with. Since these products are primarily destined for the export market, U.S. hardwood exporters will not be penalized by the existing tariff on lumber imports.

Expansion of the market for softwood lumber and logs will require a substantial effort on the part of producers, industry trade associations, and government agencies. Specific efforts designed to promote the construction of Western style platform houses and the use of low-rise construction techniques should be carefully considered. Other efforts designed to promote the use of softwood products in applications where tropical hardwoods command a large portion of the market should have top priority. Governmental efforts to reduce tariffs on sawnwood products should also proceed concurrently with these promotional efforts.

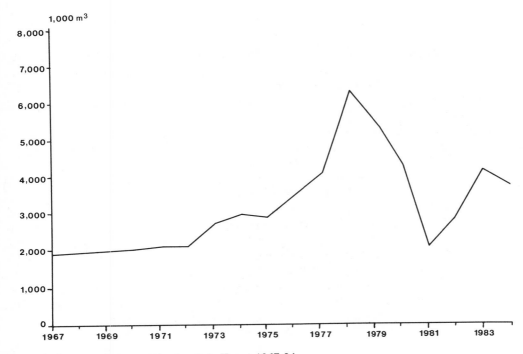

Figure 1. Sawnwood consumption levels in Korea, 1967-84.

Specific market opportunities include the promotion of Douglas-fir and ponderosa pine joinery products. Industrial applications such as subway construction also appear to be an area in which large gains in market share are possible.

Expansion of the market for temperate hardwood will also require a substantial effort on the part of the industry and others. Hardwood producers attempting to take advantage of this market will have to increase the Korean manufacturer's knowledge of many U.S. species. Korean producers are generally familiar with the more expensive temperate species such as walnut, oak, and maple, but are generally not familiar with other less expensive but commonly used temperate species, such as yellow poplar (*Liriodendron tulipifera*). Promotional efforts designed to increase the Korean furniture industry's knowledge of appearance, physical properties, mechanical properties, and the machining, finishing, and staining properties of both North American hardwoods and softwoods could result in the opening of new trade opportunities. An example of the potential return to exporters from such promotional efforts is provided by a recently published article (Govett et al. 1987). An exporter in the western United States is currently shipping Oregon white oak (*Quercus garryana*), Oregon or bigleaf maple (*Acer macrophyllum*), and Oregon ash (*Fraxinus latifolia*) to Korea for use in furniture manufacturing. This trade was the result of the exporter's efforts to find a market opportunity for western hardwood species by locating a specific company to approach that would be a likely market leader in the use of such species. Representative samples of wood in various forms were provided for the potential customer's inspection and testing. These efforts have resulted in shipments of container loads of logs and orders for follow-on shipments of more logs and lumber.

Several of the smaller segments of the industrial sector should not be ignored by either softwood or hardwood producers. North American hardwoods and softwoods appear well suited for use in the manufacture of truck flooring. Use of naturally durable hardwoods or treated softwoods in the construction of vinyl houses may also offer a market opportunity for certain U.S. producers.

SUMMARY AND CONCLUSIONS

Attractive opportunities exist for American producers interested in serving the Korean market. With Korea's growing economy and increasing standard of living, it is only logical to expect demand for hardwoods and softwoods to expand. The potential opportunities available to U.S. producers are further enhanced by the declining supply of tropical hardwood logs and Korea's own limited timber resource.

Although estimates of future market growth vary widely, a relatively conservative projection suggests that U.S. exports of solid-wood products could total 1.88 million m^3 by the year 2000, assuming a moderately expanding Korean economy and that U.S. producers can expand their share of market to 40%. However, even if U.S. producers are not able to increase their market share significantly, we can expect to see a substantial demand for U.S. hardwoods and softwoods for the remainder of the century.

The greatest opportunities for softwood producers in the near term appear to lie in increasing log exports to Korea. In the longer term, promotional activities designed to encourage the use of Western style construction techniques and the use of softwood species in applications currently dominated by tropical hardwoods appear to hold the greatest promise. Similarly, the furniture and musical instrument industries appear to hold numerous opportunities for U.S. hardwood producers in both the short run and the long run, although promotional efforts to increase the Korean manufacturer's knowledge of certain species will be required to increase sales of many temperate species.

REFERENCES

Anonymous. 1986. Volume of forestry exports grew by 14%. Chilean Forestry News. 7 p.

Blatner, K. A., R. L. Govett, and Wae-Jung Kim. 1987. A profile of the Korean market for softwood logs and lumber. West. J. App. For. 2(1):17-20.

FRI. 1982. An illustration of forest products utilization, pp. 693-744. Forest Research Institute, Seoul, Korea. Translated from Korean by Wae-Jung Kim.

Govett, R. L., K. A. Blatner, and Wae-Jung Kim. 1987. The Korean market: Opportunities for American hardwood exporters. For. Prod. J. In press.

NFPA. 1985. Building the Korean market: A $200 million potential by 1990. International Department White Paper. National Forest Products Association, Washington, D.C. 7 p.

———. 1987. International trade report. National Forest Products Association, Washington, D.C. February 1987. 12 p.

Schreuder, G. F. 1986. Productive forest resources management: Forest product marketing. Field Doc. ROK/82/013. Food and Agriculture Organization of the United Nations, Seoul, Korea.

USDA. 1987. Wood products: International trade and foreign markets. USDA Foreign Agriculture Service. FAS Circular WP1-87. Washington, D.C. 40 p.

Chapter 8

U.S. Hardwood Trade in the Pacific Rim

PHILIP A. ARAMAN

The Pacific Rim, or the Northwest Pacific Rim marketing area, has become an important trading area for U.S. hardwood products. For the last five years, shippers from all hardwood producing areas (East, West, South, Appalachia, and the North) have increased exports to Pacific Rim countries. Products shipped in 1985 were mainly logs, lumber, and veneer, as shown in Table 1; and the major importers were Japan, Taiwan, and Korea. In Japan, U.S. hardwoods are used as substitutes for domestic hardwoods in fine hardwood products (mostly furniture). In Taiwan and Korea, U.S. hardwoods are used in furniture for export to the United States, Japan, Canada, and Europe. In general, U.S. hardwood is competing with Southeast Asian hardwood products in Japan and Taiwan. As hardwood supplies from Southeast Asia continue to tighten, more U.S. hardwoods may be purchased to fill material voids for furniture production.

The goals of this paper are to explain the differences between U.S. domestic and export hardwood products and to discuss the shift and growth of export markets. In addition, the emergence of the Pacific Rim market will be documented; and the demands of the major customers, Japan and Taiwan, will be examined. Potential export dimension products for Japan, Taiwan, and Korea will also be discussed.

HARDWOOD EXPORTS

Specifications and preparations for export logs, lumber, and veneer are often different from those for the domestic market. Log exports are primarily veneer quality. Export veneer is cut thinner and clipped and packaged differently than in the United States.

The differences are greatest for hardwood export lumber. It is sold kiln dried, end coated, branded, labeled, and strapped with substantial bands and protective corner cardboard. The lumber is usually reinspected after kiln drying to remove poorly dried boards and is sometimes completely wrapped with plastic for protection. Most bundles are accompanied by a tally sheet showing the measurements of all boards. In the past, none of this was done for domestically shipped hardwood lumber except for kiln drying. Now some of the more progressive companies are shipping their domestic lumber in the same manner as export lumber.

Shipping is also quite different. Most export lumber is shipped in protective containers that provide much needed security, whereas most domestic lumber shipments are made on flat beds and with tarps when kiln-dried material is being shipped.

HARDWOOD MARKETS

In 1975, Pacific Rim demands for U.S. hardwood logs, lumber (including some rough dimension), and veneer were minor (Table 2). Canada dominated by purchasing over 71% of U.S. exports, Europe was second with almost 25%, leaving the Pacific Rim with under 4%. Smaller amounts have been and are shipped to other parts of the world. By 1980, strong demands from

Table 1. U.S. hardwood product exports to Pacific Rim countries in 1985.

Country	Logs Quantity (MBF)	Logs Value ($1,000)	Lumber Quantity (MBF)	Lumber Value ($1,000)	Veneer (1/8 inch) Quantity (M Sq Ft)	Veneer (1/8 inch) Value ($1,000)	Total Value ($1,000)
Australia	54	24	4,921	1,501	2,683	315	1,840
China	59	15	166	107	368	39	161
Hong Kong	505	399	2,650	1,391	1,739	138	1,928
Japan	5,218	8,165	43,372	29,379	12,046	608	38,152
South Korea	3,062	3,625	3,851	2,630	19,923	2,285	8,540
New Zealand	8	21	231	200	470	114	335
Pacific Islands	--	--	48	37	--	--	37
Philippines	--	--	98	105	--	--	105
Singapore	554	603	741	688	8,029	724	2,015
Taiwan	18,473	15,510	46,326	37,459	76,250	6,005	58,974
Total	27,933	28,362	102,404	73,497	121,508	10,228	112,087

Country	Plywood Value ($1,000)	Railroad Ties Value ($1,000)	Siding Value ($1,000)	Flooring Value ($1,000)	Wood Blocks, Blanks, Sticks Value ($1,000)	Total Hardwood Value ($1,000)
Australia	9	--	4	5	65	1,923
China	--	--	--	--	--	161
Hong Kong	--	--	--	83	26	2,037
Japan	175	--	--	2,511	2,218	43,056
South Korea	4	--	--	375	1	8,920
New Zealand	--	--	6	--	6	347
Pacific Islands	85	--	--	--	--	122
Philippines	10	--	--	--	--	115
Singapore	--	--	--	24	5	2,044
Taiwan	65	--	--	--	242	59,281
Total	348	--	10	2,998	2,563	118,006

Source: U.S. Department of Commerce.

Table 2. U.S. hardwood (log, lumber, and veneer) exports for 1975, 1980, 1985, and 1986 (in million board feet).

Destination	1975	1980	1985	1986*
Canada	147	202	168	176
Europe	51	256	165	249
Pacific Rim	8	33	131	208
Total	206	491	464	633

* Estimated.

Europe raised it to first place, with 52% of U.S. exports, while the Canadian share dropped to 41%, and the Pacific Rim share increased to almost 7%. Changes continued, and 1985 figures show the emergence of the Pacific Rim as a full partner, demanding 28% of U.S. hardwood log, lumber, and veneer exports, while European and Canadian demands both dropped to 36%. Estimates for 1986 show even more change, with exports to all three markets increasing. The European share increased to 39%, and the Pacific Rim share increased to 33%, while the Canadian share dropped to 28%.

Why have hardwood exports to the Pacific Rim been increasing so fast during the 1980s? Or, why have our hardwood producers turned to this market to sell their products, and why have they been so successful in the last five years? Several factors can be credited.

During this period, American hardwood sales in Europe and Canada were hurt by the increased value of the U.S. dollar and the slowed economic growth in these countries. At the same time, U.S. imports of furniture and furniture parts were increasing, thus reducing domestic demand for U.S. hardwood logs, lumber, dimension, and veneer. On the other side of the globe, log embargoes by Indonesia, Malaysia, Singapore, and the Philippines and cheaper plywood production in some of these nations caused restructuring and redirection of industries from plywood to furniture in Taiwan and Korea. This created needs for different materials, such as our temperate hardwoods. Japan's hardwood consumption was putting pressure on its limited hardwood resources, prompting the Japanese to turn to America, where the hardwoods are similar in many respects to those of Japan and are readily available.

Because of their dominance as customers for U.S. hardwoods, Japan and Taiwan will be looked at more closely in the following sections.

Exports to Japan

Japanese imports of American hardwoods have been mainly in the form of lumber (Table 3). On a value basis, 1985 exports to Japan totaled $38.2 million. Of this total, 77% was lumber, 21% logs, and 2% veneer. On a quantity basis, 89% was lumber, 11% logs, and less than 1% veneer. In 1986, the percentages were 85, 14, and 1.

Unlike other major hardwood lumber export markets, the Japanese market is not dominated by oak. It accounts for less than one-fifth of these exports. Seven out of ten of U.S. exports are listed

Table 3. U.S. log, lumber, and veneer exports to Japan in 1985.

Species	Logs		Lumber		Veneer	
	Quantity (MBF)	Value ($1,000)	Quantity (MBF)	Value ($1,000)	Quantity (M Sq F)	Value ($1,000)
Birch	10	13	--*	--*	858	17
Maple	274	137	3,784	1,334	106	24
Red oak	180	135	6,148	5,033	7,843	448
White oak	636	646	1,370	1,096	228	29
Ash/hickory	--*	--*	558	490	--*	--*
Walnut	617	1,744	1,312	1,291	133	50
Other	3,501	5,490	30,200	20,135	283	40
Total	5,218	8,165	43,372	29,379	9,451	608
1986 Totals	10,024	NA	61,162	NA	20,413	NA

Source: U.S. Department of Commerce.
* Listed in "other" category.

in the "other" category, and 68% of this lumber is shipped in dressed form. Most of this material appears to be red alder produced in the Pacific Northwest. The remainder of the "other" category wood is predominately yellow poplar, black cherry, and cottonwood. Information showing the exact species and quantities by species is not available.

Most of these exports are destined for use in furniture or cabinet products, as showings at the recent 1985 Tokyo Furniture Show confirmed. Fine-grained hardwoods and oak were the predominant woods, but the oak appeared to be mostly Japanese white oak. Japanese industry people stated they need and use U.S. oak for legs and trim work, and they use the more expensive and preferred Japanese oak for the larger, more visible surfaces.

The Japanese furniture manufacturers using oak prefer and pay more for quarter-sawn lumber or quarter-sliced veneer with tight growth rings. In the production of domestic oak material in Japan, producers cut and separate quarter- and flat-sawn material. The material is then sold separately with the quarter sawn obtaining a premium.

Japanese furniture manufacturers also like to use uniform light colored, fine-grained hardwoods that can be given a natural finish or stained. Several U.S. hardwoods meeting these criteria that were seen at the Tokyo Furniture Show included red alder, black cherry, cottonwood, maple, and sap-one-side yellow poplar. Most of these hardwoods are being used as substitutes (Table 4) for

Table 4. Some U.S. hardwoods used in place of Japanese hardwoods.

Japanese Hardwood	American Hardwood Substitute
Kaba (Japanese birch)	Red alder
Nara (Japanese oak)	Red or white oak
Shina (basswood)	Yellow poplar or cottonwood
Sen or tamo (white ash)	White ash
Kurumi (walnut)	Black walnut

Japanese hardwoods, which are either in short supply or are more expensive to purchase. U.S. promotional efforts are also introducing additional hardwoods such as gum, aspen, and soft maple.

Exports to Taiwan

In 1985 almost $59 million worth of hardwood logs, lumber, and veneer were shipped to Taiwan from U.S. ports (Table 5). As with Japan, the order of importance of exports to Taiwan was lumber (64%), logs (26%), and veneer (10%). On a quantity basis, 69% was lumber, 28% logs, and 3% veneer. In 1986, the percentages were 80, 17, and 3.

Species sent to Taiwan were mainly red and white oak. On a quantity basis, 96% of the lumber, 83% of the logs, and 63% of the veneer were oak exports, with the majority being red oak. The remaining demands were for woods such as birch, ash, walnut, and cherry.

Why is there such a great demand in Taiwan for American oak? Taiwan's major furniture producers are export oriented, and the United States is the largest market in the world for furniture. Since oak furniture is in the greatest demand in the United States, Taiwan manufacturers want and need oak. Regional supplies of oak from Japan are expensive and available only in limited quantities. So, they have turned to the United States for their rapidly growing oak needs.

Observations at the recent Taiwan Furniture Show in Taipei and the fall Southern Furniture Show in High Point, North Carolina, revealed that rubberwood and other Southeast Asian species are being used in furniture parts and furniture produced in Taiwan for export. It appears, however, that the companies using a great deal of rubberwood are those selling to retailers in the United States and not those dealing with U.S. manufacturers. Most of the companies selling to U.S.

Table 5. U.S. log, lumber, and veneer exports to Taiwan in 1985.

Species	Logs Quantity (MBF)	Logs Value ($1,000)	Lumber Quantity (MBF)	Lumber Value ($1,000)	Veneer Quantity (M Sq Ft)	Veneer Value ($1,000)
Birch	6	6	--*	--*	3,205	156
Maple	300	56	517	303	--*	--*
Red oak	14,559	12,701	36,560	30,012	44,033	3,511
White oak	809	992	8,020	6,147	4,088	437
Ash/hickory	--*	--*	379	321	--*	--*
Walnut	128	385	309	246	2,443	246
Other	2,671	1,370	541	430	22,481	1,655
Total	18,473	15,510	46,326	37,459	76,250	6,005
1986 Totals	18,411	NA	88,767	NA	114,166	NA

Source: U.S. Department of Commerce.
* Listed in "other" category.

manufacturers are using U.S. hardwoods that American consumers are accustomed to seeing in furniture products.

Potential Export Dimension Products for Japan and Taiwan

During the past ten years, U.S. hardwood lumber exports to the Pacific Rim and Europe have grown significantly. During this same period, lumber exports to Canada have also increased. The overall increases have tightened domestic supplies and increased prices of certain domestic species and grades of hardwood lumber, particularly export quality—primarily Firsts and Seconds (FAS). At the same time, lower grade material has remained available at a relatively stable price.

As an alternative to export lumber, we propose the exportation of standard sizes of rough dimension (Araman 1987, Hansen and Araman 1986, 1987). The major advantage of standard-size dimension is that it can be made from the lower, more abundant grades of lumber, while providing the same clear-quality material the export market has come to expect. This type of material is currently available only from FAS lumber or from standard size rough dimension manufactured in Asian countries. In addition to abundant resources, the United States has the physical means on line to produce these export products (i.e., kiln-drying and rough-dimension production capacity).

For Japan we suggest marketing export rough dimension in 4-inch length increments starting at 12 and running up to 72 inches, and in 12-inch length increments from 84 to 108 inches. These sizes correspond with metric sizes produced and used in Japan and Europe. Widths are random, ranging from 2 inches up. Two nominal thicknesses are recommended: 1 and 1-1/4 inches.

For Taiwan and others in the Pacific Rim area making furniture for export to the United States, we recommend standard-length, random-width dimension or glued panels in sizes developed for use by American furniture manufacturers.

Although standard dimension from the United States has many resource advantages, successful introduction of standard-size dimension on the export market will depend largely on competitive advantage. The cited publications discuss several factors likely to affect market acceptance.

SUMMARY

In 1985, of the U.S. hardwood log, lumber, and veneer exports, 28% was shipped to the Pacific Rim market, primarily Japan and Taiwan. Taiwan manufacturers are purchasing mainly red and white oak to process into finished parts and furniture for the export market. Their largest furniture export market is the United States. According to Taiwan statistics, 63% of their wood furniture exports from 1980 to 1984 were shipped to the United States.

Japanese purchasers are buying mostly the "other" species, such as red alder, black cherry, yellow poplar, and cottonwood. Two-thirds of these purchases are dressed or planed, kiln-dried lumber. In contrast with Taiwan, the U.S. hardwoods going to Japan are used as substitutes for their hardwoods in the production of furniture for domestic sales.

It is clear that the Pacific Rim has become a major market for hardwood from the United States. We believe ample supplies of hardwood exist to meet future increases in demand, provided ways can be found to accommodate a greater variety of grades and species. Standard-size rough dimension could help with the grade utilization situation by providing a means to make unexportable lumber into an exportable product. Tightening of hardwood supplies from Southeast Asia should help provide opportunities for a greater variety of American species in the Pacific Rim area.

REFERENCES

Araman, P. A. 1987. Standard sizes for rough-dimension exports to Europe and Japan. Wood and Wood Products 92(6):84-86.

Hansen B. G., and P. A. Araman. 1986. Hardwood blanks expand export opportunities. World Wood 27(1):34-36.

———. 1987. A look at potential price ranges for red oak rough dimension exports in standard sizes. Wood and Wood Products. In press.

Chapter 9

The Market for Softwood Lumber and Plywood in the People's Republic of China

STEVE LOVETT and ALBERTO GOETZL

A recent study commissioned by the National Forest Products Association (NFPA) asserts that the People's Republic of China (PRC) could become the single largest importer of U.S. wood products in the next five years (NFPA 1986). The study, sponsored by the Department of Agriculture's Foreign Agricultural Service (FAS) and conducted by DBC Associates in Beijing, China, provides a comprehensive analysis of the Chinese wood market, resource conditions, distribution systems, and manufacturing. The purpose of this article is to highlight the major findings of the study.

BACKGROUND

China is a vast area characterized by an agrarian economy, massive population, limited natural resources, and an all-pervasive political structure. The country is only slightly larger than the United States, but has five times as many people. The large population and a rapidly developing economy have created an enormous demand for housing and other construction. Rapid growth has been accommodated through significant economic reforms under the current leader, Deng Xiaoping, but China is not moving toward capitalism. It is a socialist country with a totalitarian government, run under state planning in all industrial sectors. Economic reforms during the early 1980s introduced incentives geared toward increasing production in the industrial and agricultural sectors. The reforms have not resulted in a large free market readily accessible to U.S. producers, although China's developing economy has created enormous demand which can only be met by imports.

TIMBER RESOURCES

Forest land in China covers roughly 99 million hectares (227 million acres), or 12% of the total land area; 73% of the forest land can be considered commercial (available to supply industrial roundwood). However, most of the forest land is located in remote areas of the northeastern and southwestern regions. Accessible forests are generally poorly stocked and overexploited. Annual harvests, estimated to be on the order of 64 billion board feet (290 million m^3), appear to exceed annual growth by a worrisome 32%. In some provinces, removals exceed growth by even larger percentages. The proportion of softwoods in the inventory has also been steadily declining, from about 70% in 1965 to 58% in 1984 (Figure 1). It is important to note that there is timber harvested within the context of state plans and there is timber that is harvested outside of the authorized program. About a quarter of the harvest is for planned consumption in state-approved projects (14.7 billion board feet or 67 million m^3). The balance is cut for firewood, agricultural applications, and peasant home construction. Much of the harvest goes unreported; much is lost through fires and logging inefficiencies (Figure 2).

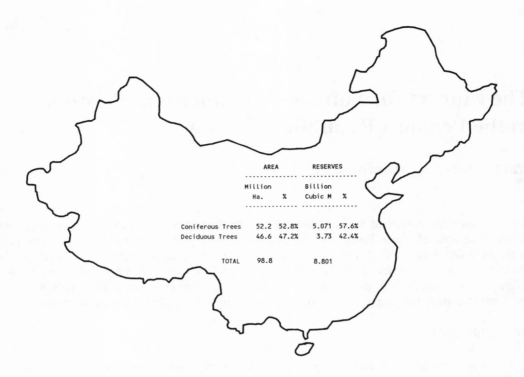

	AREA		RESERVES	
	Million Ha.	%	Billion Cubic M	%
Coniferous Trees	52.2	52.8%	5.071	57.6%
Deciduous Trees	46.6	47.2%	3.73	42.4%
TOTAL	98.8		8.801	

Figure 1. China's softwood and hardwood reserves. Only 12% of China's vast land area is forested; 58% is of coniferous species. Source: *Contemporary China's Forestry* (1985).

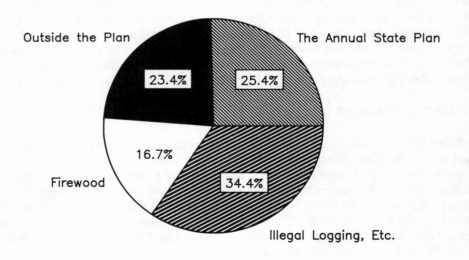

Total = 299 Million Cubic Meters

Figure 2. Total wood consumption in China. Planned consumption accounts for only a quarter of the total.

To prevent a timber famine, the Ministry of Forestry has been charged with an ambitious afforestation program. The program has successfully increased the land devoted to forests during the past decade, but not enough to meet anticipated future needs. The forest plan is to increase forest cover to 20% of the total land base by the year 2000 and follow a nondeclining, even-flow, sustained yield management policy. But illegal logging hinders these objectives.

The state has made significant changes in its forest and wood policies in recent years. For example, Marxist philosophy places no economic value on natural resources until labor is added. Therefore, standing trees have no economic value until harvested and processed. This has resulted in overexploitation of the forests, particularly in the rural cooperatives, since timber has been viewed as a free commodity. Reforms in the early 1980s brought a revamping of this system. Standing trees are now valued economically, reflecting to some extent their replacement value. Inclusion of raw material costs in the final prices of wood products is now also encouraged.

China has a goal of producing over 22 billion board feet (100 million m^3) of timber inside the state plan by the year 2000, nearly double its current production rates. Because of the host of problems associated with the country's domestic resources, it is highly unlikely this goal will be realized.

The official policy of the government discourages the use of wood products in order to preserve domestic timber supplies. Use of concrete, plastics, steel, and cardboard are encouraged where wood might otherwise be used. But numerous problems have arisen through this policy and most users question its effectiveness.

CHINA'S WOOD INDUSTRY

The Chinese wood products industry is dominated by small sawmills. About 155 large mills are operated by the Ministry of Forestry; another 400 to 600 large mills are operated by provincial or municipal timber corporations and ministries. Thousands of other small sawmills dot the country. The mills are generally poorly designed and inefficient. Refurbishment and new construction of sawmills do not appear to have a high priority in state planning.

Elements of Demand

Consumption of wood products in China can be grouped into four categories: (1) consumption inside the state plan, (2) consumption outside the plan, (3) firewood, and (4) miscellaneous uses and losses.

Consumption inside the state plan includes that which is procured by the ministries, industrial enterprises, and urban construction authorities. The Ministry of Forestry is the primary domestic supplier; allocation and distribution are handled through the China Timber Corporation and the provincial timber corporations. Supplies of wood to all processors and users are approved by the state in accordance with needs set forth by provincial and state planning commissions. Table 1 outlines consumption of wood within the state plan, by market sector.

Consumption outside the plan is permissible by collectives and individuals, provided the harvest does not exceed growth. Local forestry bureaus are responsible for enforcement. Timber harvested by collectives and individuals can be sold to the state to meet its supply requirements, but most is used by peasants and collectives for housing, rural construction, furniture, and so forth.

A large amount of roundwood is burned for fuel each year. This use is approved, but not policed. The equivalent of 24 million cords (50 million m^3) of wood is believed to be burned annually. Finally, a dramatic amount of wood—estimated at 23 billion board feet (103 million m^3)—is consumed illegally or is wasted through destruction or inefficiencies.

Wood consumption has increased with the rise of the Chinese equivalent to the gross national product, referred to as the gross value of industrial and agricultural output (GVIAO). As this measure has grown from 224 billion renminbi (RMB) in 1965 to 1,327 billion RMB in 1985,

Table 1. Consumption of state-planned domestic wood, by market sector.

Market Sector	%	Quantity (million m^3)		
		1981	1984	1985
Capital construction Doors, windows (50%) Concrete forming (45%) Scaffolding (5%)	20.2%	10.0	12.9	13.5
Mining props	13.6	6.7	8.7	9.1
Packaging	10.1	5.0	6.4	6.7
Fuelwood	8.1	4.0	5.2	5.4
Paper	7.5	3.7	4.8	5.1
Furniture	4.7	2.3	3.0	3.1
Agricultural uses	3.4	1.7	2.2	2.3
Rolling stock	3.0	1.5	1.9	2.0
Plywood	1.4	0.7	1.0	1.0
Railroad ties	1.0	0.5	0.7	0.7
Utility poles	1.0	0.5	0.7	0.7
Posts	0.8	0.4	0.5	0.5
Thinning of forests	5.9	2.9	4.0	4.2
Other	19.2	9.5	11.8	12.7
Total	100.0	49.4	63.8	67.0

Source: Ministry of Forestry, Forestry Economics Research Agency (n.d.).

planned wood consumption has nearly doubled from 9 billion board feet (41 million m^3) to 17 billion board feet (76 million m^3), including imports (Figure 3). Wood consumption per unit of the GVIAO, however, has declined. China's planners have determined that the minimum level of roundwood consumption needed to meet economic objectives is .08 million m^3 per billion RMB of growth. This is about the current level. At the same time they have determined that wood consumption must drop to .056 million m^3 per billion RMB in order to conserve domestic resources. Presumably the difference is to be made up by imports.

Even using the most conservative estimate of wood use leads to the conclusion that the growth in wood demand under the state plan during the next twenty years will be dramatic. It is projected to increase from 17 billion board feet (76 million m^3) to 35 billion board feet (156 million m^3) by the year 2000. Even if planned consumption should fall short of that projection, the need for increased imports is anticipated (Figure 4). China's planners believe that at least 13% of planned consumption will consist of imports. This is probably unrealistically low since imports currently account for 12% of consumption.

More Converted Products Likely

The maximum annual sawmilling capacity in China today is estimated to be only 25 million m^3 (10.5 billion board feet lumber tally). With refurbishment and new construction, this capacity could increase to 30 million m^3 (12.5 billion board feet) by 1990. This capacity is not likely to be exceeded because there are few existing plans to improve sawmilling capacity or to make other investments in existing mills. And no increase in the growth rate of lumber production is planned for the next five years. Moreover, the lion's share of the sawmilling capacity for imported logs is located in the coastal provinces where few opportunities exist for expansion. Imported logs can be converted at these sites, but as the capacity is exceeded, imported lumber will be in demand. There-

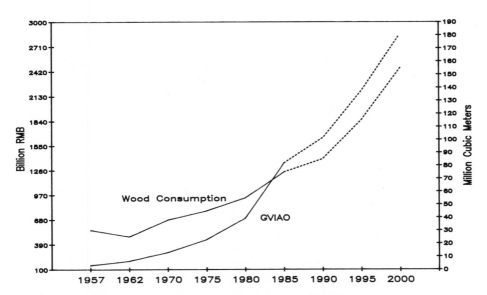

Figure 3. Forecast of wood consumption based on projections of GVIAO.

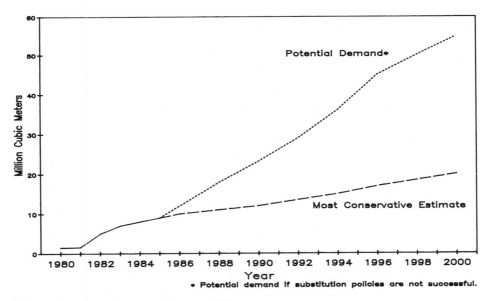

Figure 4. Import projections through the year 2000.

fore, the study concludes that with an aggressive marketing strategy in place, the United States could capture 55% of total lumber imports, 650 million board feet (1.5 million m³).

The United States faces stiff competition with the U.S.S.R., Canada, and Chile in supplying the Chinese market. During the first seven months of 1986, U.S. exports to China were down 38%. This is attributable to a shortage of foreign exchange. In 1987 the U.S. share is expected to return to the 52% level it has enjoyed for several years.

China first began importing softwood lumber in 1981. Although the volume was small, 63 million board feet, the United States supplied 55%. In 1985, total lumber imports represented 122 million board feet, but the U.S. share declined to 14%. Canada has become the major supplier of lumber to China.

The leading supplier of plywood to China is Indonesia, but the Philippines and Taiwan are also suppliers. Although of low quality and strength, the imported plywood appears sufficient for the Chinese furniture industry. Plywood has not yet found a foothold in construction. A negligible amount of softwood plywood is currently being imported.

CHINA'S IMPORT POLICY

China's import decisions are largely political, motivated by the need to supply the materials for modernization in industry and agriculture, and to slow depletion of domestic reserves of timber. China imports wood for four basic reasons: (1) to make up for the domestic shortage of wood, (2) to obtain large size and high quality wood, (3) to balance the available species, and (4) to supply improved exporting enterprises.

Imports of wood must be approved by the State Planning Commission and must conform to national priorities in the state plan. User demand is of secondary importance when import decisions are made—and is insignificant relative to state development priorities—but is important for influencing long-range market development.

State policy dictates that reserves of foreign exchange be used judiciously, so barter trade is preferred to expenditures whenever possible. The principal factor determining source of imports is price. All requisitions to import wood products from North America are consolidated under the China Timber Importing and Exporting Corporation (CTIEC) for purchase approval.

The Competition

Because of its proximity, the U.S.S.R. can be a major supplier of roundwood to China. In fact, unconfirmed reports of an agreement on timber trade between the two countries have circulated. Since the agreement would provide for barter trade of unspecified commodities, it would require no use of foreign exchange. The purported agreement would seek to increase Soviet imports of softwood roundwood to 10 million m^3 per year by 1991, a 500% increase from current levels. Most sources doubt this can be accomplished. In 1985, the U.S.S.R. was contracted to supply 2.2 million m^3 (924 million BF); only 1.8 million m^3 (756 million BF) were actually imported. The Chinese have considered sending their own crews into the U.S.S.R. to harvest timber, but currently this is not a likely prospect.

Canada probably represents the major competitor for the United States in China. Canada has increased its exports of softwood lumber to China, accounting for over 85% of the import demand. China and Canada have embarked on a number of agreements to improve their trade relationship in recent years, and the Canadian wood products industry has been aggressively marketing in the country. The Canadian forest product industry is opening an office in China, and recently signed a joint venture to build a model farm there. Finally, Chile is supplying about 7% of China's imports of roundwood.

As of July 1986, CTIEC had greatly reduced orders for U.S. logs. This can be attributed to the $14 billion depletion of foreign exchange reserves incurred in 1985, hence a shortage of foreign currency. Imports are now primarily through countertrade agreements, with the U.S.S.R. and possibly Chile. Assuming the economy improves in 1987, imports from the United States will recover. Wood will increasingly be supplied by parties open to creative trade deals.

Tariffs are imposed on wood imports to a varying degree, depending on the level of conversion. They range from 3% for rough sawn logs and cants to 9% for rough green lumber and 40% if planed or otherwise finished. Most veneers and plywood have tariffs of 12 to 30%. The tariffs are geared to encourage importation of raw material rather than finished goods, thereby adding the value domestically. In addition, unpublished product taxes imposed by the Ministry of Finance average about 10% of all forest product imports. Coupled with the published rates, these taxes make the total tariff obligation among the highest in Asia—higher even than in Japan.

Market Opportunities

There is little doubt that the Chinese need for wood is huge. The primary market remains in the priority areas of China's economic development: capital construction, defense, industrial uses, and transportation. Wood use patterns differ from the United States and other countries. Construction uses of wood do not tend to be in framing, but in window and door construction, and concrete forming. Scaffolding also is made of lumber. Mining props and packaging account for the next two largest categories of consumption. Most construction is of concrete frame and brick.

Chinese peasant housing utilizes wood in windows, doors, and roof structures. This is a vast potential market because peasant homes are being built at the rate of 9 to 10 million per year.

U.S. Exports to China

China has been the fastest growing customer for U.S. wood product exports. Exports to China have increased from nearly zero in 1979 to over $328 million in 1985, made up almost entirely of softwood logs, with small volumes of lumber and other products. Exports have fallen dramatically in 1986 in response to the retrenchment by China's leaders (Figure 5). Foreign exchange has been in short supply and therefore unavailable for import purchases. This is not believed to be indicative of future trends.

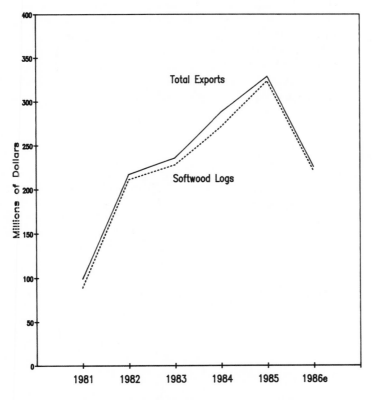

Figure 5. Value of U.S. wood exports to China, 1981-86. Source: U.S. Department of Commerce; FAS.

U.S. Industry Efforts

The NFPA study recommends a three-pronged marketing strategy for U.S. industry, with over twenty specific actions. Success in furthering exports to the People's Republic of China will require working within the highly bureaucratic and politically motivated Chinese system. Three overall objectives were put forth in the study.

1. Bring about the removal of tariff and nontariff barriers to the importation of softwood lumber and plywood. The U.S. industry must seek lower tariffs, particularly for finished products, in part by demonstrating the importance of importing these products in meeting state objectives. By educating the Chinese about uses and qualities of U.S. products, we can attempt to influence the use of wood in more applications and show the economic and quality advantages of U.S. products.

2. Target geographic and product markets for acceptance and development of converted products. This includes developing and promoting sizes and standards currently acceptable to the Chinese. Efforts to this end will be most successful if it can be shown to provincial timber corporations that imports of converted products can mean higher profits under China's limited plan for profit potential.

3. Influence wood product acceptance through cooperative technical efforts. The current lack of product standards and the need for technical assistance provide an opportunity to introduce the Chinese to U.S. sizes, grades, codes, designs, and engineered structures. Once again, it is critical to work within the Chinese system. For example, building designs should incorporate indigenous materials as well as imported products.

A Special Challenge

Inroads into China will not come easily or quickly for U.S. industry. Because of the inherently political nature of China's industry, time must be taken to build and maintain relationships, the quality and honesty of any relationships being critical. By being patient and adaptable to the Chinese market, the U.S. wood products industry can look forward to a mutually profitable and long-term trade relationship. The NFPA study on China quantifies what has been suspected for the past few years: if a sustained strategy, unifying the industry, government, and trade associations were developed, U.S. producers could find a growing market for logs, lumber, and panels in the People's Republic of China.

REFERENCES

Contemporary China's Forestry. 1985. Edited by Li Ting. Ministry of Forestry, Beijing.

Ministry of Forestry, Forestry Economics Research Agency. n.d. China's forest products industry: A report on problems of supply and demand. (Translated by DBC Associates.) Ministry of Forestry, Beijing.

National Forest Products Association. 1986. The market for softwood lumber and plywood in the People's Republic of China. 2 vols. NFPA, Washington, D.C.

Chapter 10

The Outlook for Timber Demand in the People's Republic of China

JIA JU GAO

PRODUCTION, SUPPLY, AND DEMAND FOR WOOD

China is a densely populated country with highly insufficient forest resources. The growing stock per capita is only about 10 m^3, which is over eight times less than the world average. The disparity in timber consumption is even greater: at 0.05 m^3 per capita per year, the consumption of wood for industrial purposes is about thirteen times lower than the world average. The development of the national economy requires ever larger supplies of industrial wood. This has pushed up the level of the annual cut, with adverse effects on the environment and on the sustained yield of the forests. But the foundation of sound forest management is the principle of sustained yield (the total annual cut in a designated area is limited to the volume that can be replaced each year by new growth). This is one way of ensuring adequate future supplies. We believe that the reforestation program in China is the most ambitious any country has undertaken; however, the government has decided to stabilize the annual cut at its present level of 50 million m^3 within the state plan. In this situation, the gap between the supply of wood and the demand for forest products could become more acute every year, adversely affecting various sectors of the national economy and the living standard of the people.

An important way to alleviate this situation is to utilize more fully the available supply of raw materials. There is, in fact, a growing gap between the rising demand for forest products and the limited availability of raw materials. Another way of overcoming this shortage is to import logs and lumber. The past demand and future needs for imported logs are shown in Figure 1, as reported by the China Forestry Society in 1985. In the past, four states in the western United States have exported logs to China. As estimated by the USDA, these exports are shown in Table 1.

Table 1. U.S. exports of logs to China (million board feet).

Source	1980	1981	1982	1983	1984	1985	1986
Washington	43	171	359	535	590	} 1,029.6	615.4
Oregon	45	48	174	164	255		
Northern California		--	--	9	8	14	
Alaska	--	3	--	16	--		
Total	88	222	542	723	859		

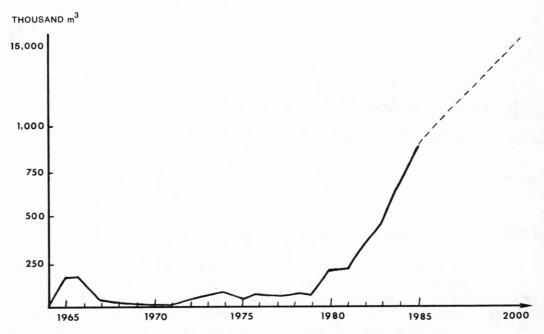

THOUSAND m^3

Figure 1. Past demand and projected future needs for imported logs in the PRC, 1965-2000.

THE CHINESE WOOD MARKET

China is a promising market for North American forest products for several reasons: (1) Since China adopted an open door policy in 1978, export trade has grown rapidly. (2) China's expanding construction industry builds 11 million m^2 of housing a year and projects 14.8 million m^2 in 1990. (3) Chinese villages now prefer to build their houses with wood roofs, trusses, and flooring. Combining wood with other building materials makes wood an even stronger, more versatile, and lower cost material. (4) Import demands have been increasing, and this trend should continue. China's demand is predicted to reach 15 to 20 million m^3 by the year 2000. To remain competitive in the international marketplace, the Chinese intend to improve business methods, showing greater flexibility than before, to meet the needs of existing customers and develop business relations with new clients.

It is disappointing that there has been a sharp fall in China's imports of U.S. logs from mid-1986 to March 1987—the so-called March earthquake among U.S. timber exporters. Quantities on order declined 50%. Since April there has been no Douglas-fir in the timber trade market near Shanghai. This is not because of changes in China's demand; on the contrary, U.S. timber always sells well in China's markets. In Jiangsu Province the sale of timber from January to April increased 195,000 m^3 over the same period in 1986, an increase of 28.6%. We bought only 771,100 m^3, less than in 1986 at the same time. We sold more than we received. We had to sell some of the timber kept in stock. In the local timber market the need for timber has increased by 30%. In the countryside, housing construction is the biggest buyer. For example, in Kun Shan county, near Shanghai, housing construction took 87% of the total wood sold for that area. Recently, the less popular hemlock has been selling well.

The reason for such a fall seems to be that our foreign exchange has not been kept in balance. Under the impact of a sharp fall in the price of petroleum because of the crisis in international oil markets, China has been forced to reduce its imports of logs from North America. But we will import more from the Soviet Union. Barter trade is of mutual benefit, for both importing and exporting. Our policy is to export as much as possible. We need foreign currency to finance our own imports of raw materials. China Timber Importing and Exporting Corporation held a wood industries products fair in October 1986 in Guangzhou and will hold one in September 1987 in Tokyo.

FUTURE OUTLOOK

Housing construction is increasing in the United States, and a large variety of furniture and wood products are needed. If China can export those products to the United States, we will create a wonder in our log and timber trade. The investment in plants to make such products is less than that needed to set up a big, comprehensive processing plant. Also in China there is sufficient labor to produce these products. Modern wood complexes take years to build, at great expense. Small plants can be built in a much shorter time and bring immediate returns. Our new policy of opening to the outside world presents unrivaled opportunities for investment and trade. The way we see it, a joint venture means primarily less taxation and more benefit for investors. The joint venture's income tax is reduced for the first five profit-making years or may even be completely exempted. What's more, all lawful rights and interests of foreign investors are protected by the Chinese law. Net profits are remitted abroad.

Many businessmen are worried that the government might nationalize or requisition their investments and property. In China we have gone out of our way to assure foreign businessmen that such things will never happen, that China is a stable economy, and we mean what we say. Any investment involves risks, but the chance of success in China is very high. This is because China's political situation is stable and the open policy is firm, and also because the country now presents an excellent environment for investment. China is such a vast country with a large population. Our resources are abundant, our wages low. These are the advantages we offer. In return, we are interested in advanced technology and the transfer of technology. I think North America realizes that China's opening its door is a way of offering foreign companies wonderful opportunities. There are lower production costs and the possibility of gaining greater access to the China market. China has decentralized the economy considerably, and foreign companies can now deal directly with potential partners and customers. Such ties will undoubtedly be good for the expansion of mutual trade. A good businessman thinks not only of immediate interests but long-range ones as well. No doubt more people will come to China to seek friendship and to try their luck. Timidity never wins first place.

Technical Considerations
of Trade

The Role of Technology in Improving the Competitive Position of the U.S. Forest Products Industry

H. M. MONTREY and JAY A. JOHNSON

The approach that we will take in discussing the role of technology in trade is as follows: first, we will search for *lessons* that can be learned from the Japanese, the acknowledged leader in international trade; second, we will focus on the areas where technology can contribute to enhanced *competitiveness*; and finally, we will give *examples* of key technological changes and their impact.

LESSONS

At the 1986 joint meeting of the Industrial Research Institute (IRI) and the Japanese Techno-Economics Society (Coover 1986), twenty-five vice-presidents of research and development of U.S. firms agreed that the United States "will *not* be in a position to compete favorably with Japan unless we do something collectively different than we now do."

The reasons why the United States is in a poor position, according to H. W. Coover of Eastman-Kodak, are the following: (1) U.S. production facilities are deteriorating at a rate faster than we are improving them; (2) there is insufficient development of scientific and technical skills in our industrial work force; (3) technology is becoming available to more people more rapidly; (4) there is in the United States a deteriorating climate for innovation in our industry; (5) by failure to emphasize long-term investment, the U.S. competitive edge has been dulled; (6) there has been a preoccupation with short-term earnings on the part of investors; (7) U.S. international trade policies are less aggressive than those of other countries; (8) in most U.S. companies there is a lack of global marketing strategies; and (9) management bureaucracies have, in many cases, been slow to adapt to change. These are strong indictments against U.S. industry, particularly since they come from "one of its own."

But how does Japan apparently avoid these traps? What does Japan do to maintain its acknowledged status as the most competitive economy in the world? These are not easy questions to answer nor are the answers clean and simple. However, from a cursory examination of the literature and from personal contacts with various Japanese representatives, certain themes emerge. For example, Japanese companies are willing to try different things. They are not caught in a catch-22 trap as some U.S. companies appear to be (the company does not try new things because "We're not in that business," and they don't change their business because they would then have to try new things).

Another aspect in which Japan differs from the United States is in the area of implementing technology. Whereas the United States still leads in quality and quantity of scientific discoveries, the Japanese are unparalleled in their ability to concentrate on and adapt the technologies that lead to commercial leadership. The technology leading to commercialization, moreover, is well integrated with market needs and production requirements. Sales and production personnel are brought into research and development projects at an early stage. The research and development personnel are trained in marketing. Research is then focused on production development required to satisfy consumers' needs.

It is also interesting that many companies in Japan think of their firms in a symbolic way which clearly illustrates the Japanese attitude toward technology. The company is represented by a tree, the customer by the sun. The trunk and branches are the parts of the company that flourish in the radiance of the sun, but the foundation of the company rests on technology, the roots. Managers believe that technology is the most important means for growth.

COMPETITIVENESS

How do the marketing strategies of the U.S. forest products industry compare with the generalized Japanese approach? Although there is evidence of change, many U.S. companies are electing to continue the traditional focus on commodities. They are adopting a "low cost producer" strategy. However, not everyone can be "low cost." Witness the demise of parts of the U.S. steel industry. Moreover, other nations are on the same tack, and danger signals are emerging. For example, Aracruz (Brazil) puts wood into its market pulp mill at about one-third the cost of an average North American mill. Chile and New Zealand export increasing amounts of sawlogs and chips to markets once dominated by the United States.

For our industry to remain competitive, it must emulate Japan by starting with customers and meeting their market needs. Two important things must be done: (1) producers must go to the marketplace and find out what customers want, and (2) ideas for new products must be generated in conjunction with a thorough understanding of the market.

It is interesting to note that the people in Japan generating the majority of successful new product ideas are scientists and engineers. The research and development staffs are directly and intimately involved in the marketplace: they interact with customers, they visit trade shows, stores, and so forth, to see what people buy. They are integrated with the marketing and sales staffs of their respective firms. This approach or one like it is needed in order for the U.S. wood industry to be competitive in world markets.

In working from customers' needs back to manufacturing, it is important to focus on the *key user* of products. For example, Post-it note pads didn't receive support in 3M (a premier U.S. product development company) until the inventor sold the idea to executive secretaries. Several companies passed up the opportunity to acquire Xerox's duplication technology because the user was perceived to be the copier. The key user in this case was not the secretary who had to handle the messy carbon paper but the person receiving the copies.

Who is the key user (or customer) for lumber? Is it the distribution center? Contractor? Carpenter? Weekend do-it-yourselfer? Or is it the housewife who wants flat walls and has to manage a home made from lumber? A recent survey of the Italian market completed by one of the authors (Johnson) revealed that machining and appearance (no knots) were characteristics required for raw material used in high quality windows. Customer preferences were related to design and trouble-free operation. Yet specification of lumber from North America is based on modifications of traditional strength-related lumber grades. Producers have not made the effort to find out what the key user wants.

What, then, should be done to enhance the competitive position of the U.S. industry? Answer: emulate the Japanese and work back from the market. This will require a growing shift from a commodity manufacturing orientation to a needs-driven, value-added approach to the marketplace. Energy and effort must be focused on maintaining existing markets for wood and fiber as well as growing new markets. Success will depend on the ability to penetrate major markets where wood is little used (the nonresidential market, for instance) and to expand markets where the key user is the customer (i.e., do-it-yourselfer). In addition, products and conversion processes must be chosen which allow end-use requirements and customer needs to be met with the changing resource base that is likely to emerge. Technology has an important role to play in order to make this strategy a success.

EXAMPLES

To illustrate the role of technology in improving the competitive position of the United States in the area of wood products trade, a number of important technological changes affecting competitiveness will be presented. This list is not comprehensive, but the topics are included to draw attention to the central theme developed in this paper.

1. *Increased componentizing and off-site manufacturing in house and building construction.* This production method requires quality "sticks and sheets," and specifications must be met by the suppliers. There is a greater opportunity to "tailor" materials, components are engineered, and connector systems become increasingly important.

2. *Solid-wood manufacturing processes managed less to maximize recovery and more to maximize product value.* More attention will be given to drying in solid-wood processing. Changes will be required in process control strategy in which algorithms for primary and secondary breakdown will combine grade and volume decisions, not just volume. Mills will be managed for overall profit instead of for green recovery and volume per man-hour ratios.

3. *Applications of "smart" (or "expert") systems or robotics that recognize value as well as volume and shape characteristics.* Lumber will be graded for specific end uses. Improvements will be made in machine stress rating (MSR). Grade yields will be improved, and grading will be done for other properties than bending stiffness and strength (i.e., MOE and MOR)—tensile strength, for example. Marketing will be required to sell these new distinguishable products.

4. *Simplification in lumber grade rule books.* There will be fewer combinations, and architects and engineers will have an easier time specifying wood. The In-Grade Test (IGT) program (Pinson 1983) will facilitate the creation of "strength classes."

5. *More combinations of wood and nonwood materials.* There will be more chemically modified wood. Thick, stable, durable, finishable, uniform, fiber-based products will arrest the rate at which plastics have substituted for wood. Composites will combine wood and inorganic binders (e.g., cement board). Metal-plate trusses will evolve and spawn other components (e.g., I-beams, box beams, and other structural components).

6. *Increasing use of computer models.* There will be more structural models (e.g., the National Forest Products Association's "Advanced Design Procedures" for floors, walls, and roofs), thermal performance models incorporating an envelope design concept, and models dealing with fire control.

7. *New applications of oriented strand technology.* Waferboard led to oriented strandboard (OSB); OSB will lead to oriented strand lumber (OSL); and OSL may lead to various shaped components in which the orientation of the strands will dictate the design.

8. *Feasibility of using surplus hardwoods in the United States.* There is a potential for increased use of hardwood in pulp and paper, and also for increased use in emerging wafer and strand products.

9. *Biotechnology.* Biochemical control of degradation could "preserve" species. Tissue culture and genetic engineering can be used to change the nature of the raw material. Biopulping technology can be pursued.

10. *Process innovations such as those in the pulp industry.* Accelerated use will be made of low quality raw material. There will be lower basis weight papers and innovations such as press drying. There will be more combinations of wood fiber and additives (fillers and organic polymers). Tailoring products will result from process changes to meet customer requirements. More use will

be made of performance-based quality control tests to monitor product reliability (e.g., the changes in shipping container specifications).

These examples point out how technological changes are having and will continue to have an impact on product quality. This in turn will obviously affect trade. Some of the developments will be hastened by changes in the resource base. This will bring timber managers and timber converters closer together. Efforts will undoubtedly be made to profitably convert wood from intensively managed softwood stands to make use of the unique attributes of juvenile and mature wood.

SUMMARY

The basic message of this paper can be stated very succinctly as follows: (1) To enhance competitiveness, the U.S. forest products industry must emulate the Japanese approach. (2) The industry must understand customer needs. (3) The industry must convert the wood resource into products that will satisfy marketplace requirements. (4) Technology will play a pivotal role in this process. It should be obvious, furthermore, that customer needs and marketplace requirements in North America are not the same, by and large, as those in other parts of the world. A number of presentations at this symposium have underscored that fact. Specific strategies must be developed for foreign markets.

REFERENCES

Coover, H. W. 1986. Programmed innovation: Key to technological leadership. *In* R&D management for high-tech in new and mature industries. Japan Techno-Economics Society, proceedings of the October 2-3, 1986 meeting, Tokyo.

Pinson, N. 1983. Changes in material resistance evaluation: Lumber. *In* Wall and floor systems: Design and performance of light frame structures. Proceedings 7317. Forest Products Research Society, Madison, Wisconsin.

Chapter 12

Conversion Factors in Forest Products Trade: Separating Fact from Fiction

DAVID G. BRIGGS and W. RAMSAY SMITH

Worldwide there is great diversity in measuring forest products. Some products are measured in terms of volume, some in terms of area, and some in terms of weight. Many different methods have evolved in various countries for obtaining basic measurements and applying them to a specific product. Those who are in trade as a business or as an analyst or policy maker must face the inevitable problem of developing a conversion factor to express quantities used in one country in the terms of another. Furthermore, it is sometimes desirable to place all products on a comparable measurement basis such as weight. Given all of the variations in products, customary measurement methods, species, and processing techniques, the development of appropriate conversion factors becomes a complex undertaking. Oversimplification or failing to understand all the elements can lead to large conversion errors. This has been noted as a possible source of inconsistencies in reports of trade statistics by trading partners (Durst et al. 1986, Darr 1984).

How important are such conversion errors? Darr (1984) noted that disagreement on a conversion factor from Scribner to cubic meters was an issue in the softwood lumber trade controversy between the United States and Canada. He also illustrated how conversion factors add confusion to the trade statistics between the United States and Japan. Two additional examples illustrate the magnitude of conversion factor problems. Figure 1 presents Douglas-fir log exports to Japan from Washington and Oregon showing the original Scribner values (solid line) and the values converted to cubic meters. The solid cubic meter line represents a conversion of 4 m^3/MBF (cubic meters per thousand board feet), about 7.1 BF/CF (board feet per cubic foot), officially used by Japan's Ministry of Finance. The dotted cubic meter line represents a conversion based on 5.5 BF/CF (5.15 m^3/MBF) derived from the U.S. Department of Commerce Schedule B Export Regulations. The difference between these conversions is substantial, amounting to more than 20%. Pity the individual who finds the Japanese cubic meter series and, unaware of their conversion, converts to Scribner using 5.5 BF/CF or some other value. The difference between the original and reconverted quantities (dotted Scribner line) is a conversion error that will probably have the analyst perplexed.

While this example may appear contrived, it was used to illustrate how easy it is to get off the track. Situations such as this may exist in published statistics but are generally much more difficult to document; one may suspect a conversion effect but it is rarely possible to define it. One case that it has been possible to document is illustrated by Douglas-fir log export statistics (Figure 2) from Canada. (This example was brought to our attention by Don Flora, forest economist, USFS Pacific Northwest Research Station.) Since 1977, log exports have been reported in both board feet and cubic meters. If the ratio of cubic meters per MBF is examined, one discovers the use of 2.36 m^3/MBF (12 BF/CF) through 1983. Subsequently, the conversion has been 4.53m^3/MBF (6.25 BF/CF). The latter is certainly a more realistic conversion for logs, as we will discuss later in the paper. The former may have been based on the mistaken notion that 12 board feet of log scale equates to a cubic foot of log, a situation that applies only to the Brererton log rule, which was not

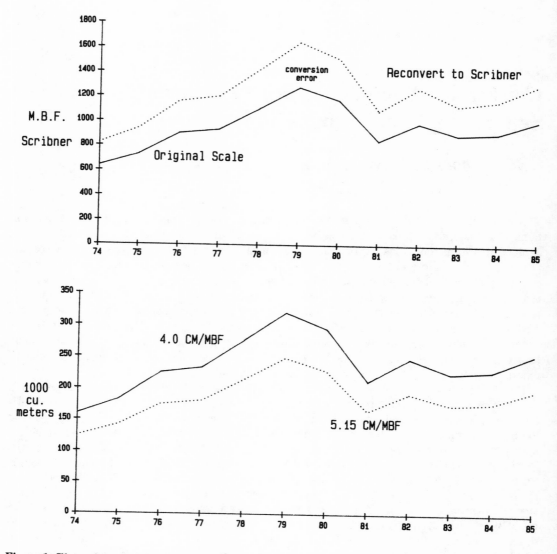

Figure 1. Illustration of conversion factor differences in Douglas-fir log exports to Japan from Washington and Oregon (1974-85) in original Scribner values (top, solid line) and values converted to cubic meters (bottom, solid line). The solid cubic meter line represents a conversion of 4 cubic meters per thousand board feet. The dotted cubic meter line represents conversion based on 5.5 BF/CF (5.15 m³/MBF). Reconversion to Scribner is shown by the dotted line in the top figure. Source: Warren (1987).

the rule used. Needless to say, this could cause problems in an analysis of trade patterns and flows, and if combined with shipment value, the implied prices could be very misleading. There are undoubtedly many other situations where conversion factor errors and changes have occurred. Many are hidden from users of the data series because of obscure or nonexistent documentation regarding whether or not a conversion was involved let alone documentation of the basis for its derivation.

This paper derives conversions of U.S. measures for logs, lumber, and chips. The aim is to illustrate methodology as well as to show potential sources of error and thereby indicate the severity

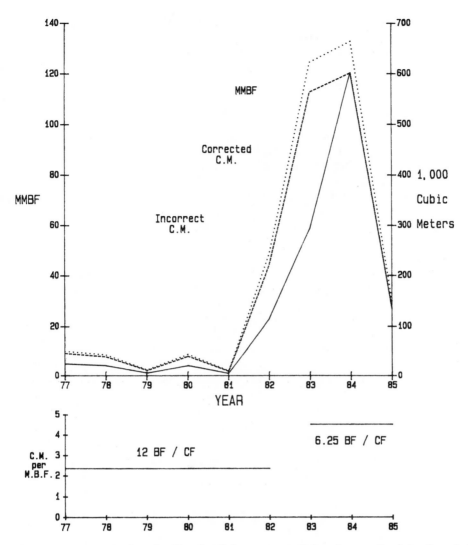

Figure 2. A conversion error in Canadian Douglas-fir log export statistics. Source: *Statistics Canada* (1977-85).

of problems for transaction or analysis purposes. A sampling procedure is offered as a solution that could be especially useful for transactions, and suggestions are made that can be valuable in interpreting statistical series.

LOG CONVERSIONS

A subtle conversion error not shown in either illustration above is the great likelihood that the true conversion factor has been varying over time because of changes in the size and taper of logs being exported. It is also common to find the same factor being applied to all species—Douglas-fir, hemlock, cedars, and so forth—even though the conversion factor will vary greatly between species because of different log size distribution and taper characteristics. Before examining some common log measurement systems and conversions, criteria for a good log rule should be established. A good log rule should (1) yield an accurate and precise estimate of the gross log volume (total fiber content), (2) provide an accurate and precise basis for estimating yields in converting the log into alternative products, (3) have the property that when a long log is cut into shorter segments, the sum of the segment volumes equals the volume of the parent log, and (4) be based on simple, easy-to-take measurements.

Scribner

The Scribner log rule is based on diagrams that predict the yield of lumber (expressed in board feet, as explained in the section on lumber conversions) based on the following assumptions: (1) lumber processing is simulated by diagramming live sawing board patterns, (2) lumber is 1 inch thick, full size, (3) sawkerf is 0.25 inch, and (4) all lumber is the full length of the log (short pieces that could be taken in taper are ignored, hence the log is treated as if it were a cylinder). Thus Scribner is a model for lumber recovery and does not really measure log volume. The assumptions in this model are obsolete compared with current manufacturing practices.

The tally of lumber, rounded to the nearest 10 board feet, forms the usual Scribner Decimal C tables. A plot of these board foot values against log diameter and length reveals that Scribner is a very nonuniform step function. These steps, and their lack of uniformity, lead to erratic and often large changes in predicted lumber recovery for very small changes in log size. Thus conversion factors based on Scribner can be volatile. Furthermore, there are differences in the way that Scribner is applied in practice. The two basic systems are the west-side or long-log system and the east-side or short-log system. These differ in whether diameter measurements are rounded or truncated, in the threshold at which segment scaling is invoked, in procedures for handling taper to estimate scaling diameters of the segments, and in procedures for treating deductions for so-called scaling defects such as rot, cracks, and so on. The key differences are summarized in Table 1. Compared with the criteria for a good log rule, Scribner fails to measure all the volume present; it is erratic, so it forms a poor basis for predicting recovery; the sum of its volumes of log segments exceeds the volume of the parent log; and it is variable in its application.

Table 1. Differences between west-side and east-side Scribner practices.

Variable	West-Side System	East-Side System
Scaling diameter	Fractional inch is truncated	Fractional inch is rounded
Length minimum for segment scaling	40 feet	20 feet
Taper assumption for estimating segment diameter	1 inch/10 feet	Taper is the difference between large- and small-end scaling diameter. (Allocate as evenly as possible to the segments, placing largest taper in segment from upper stem.)
Method for scalable defects	Diameter or length is reduced	Pie cut or squared defect is used

Cubic

Estimating cubic volume would appear to be straightforward until one realizes that several different formulas (Table 2) are used in various countries. While most of these formulas would satisfy the criteria for a good log rule, they will produce different results for the same log. In addition, any particular formula can yield different results because of variations in the measurements used. For example, measurements could be (1) actual dimensions, say to the nearest 0.1 inch, (2) west-side Scribner small-end diameter and length (small-end diameter measured in Scribner method, or taper assumption to get large-end diameter from small end), or (3) east-side Scribner small-end and large-end diameters and length.

Table 2. Examples of cubic formulas.

Formula	Cubic Foot	Cubic Meter
1. Smalian	$0.002727(d^2 + D^2)L$	$0.00003927 (d^2 + D^2)L$
2. Bruce's butt log	$0.002727(0.75d_2 + 0.25D^2)L$	$0.00003927(0.75d^2 + 0.25D^2)L$
3. Huber	$0.005454\ d_m^2 L$	$0.00007854\ d_m^2 L$
4. Two-end conic	$0.001818(d^2 + dD + D^2)L$	$0.00002618(d^2 + dD + D^2)L$
5. Newton	$0.000909(d^2 + 4d_m^2 + D^2)L$	$0.00001309(d^2 + 4d_m^2 + D^2)L$
6. Subneiloid	$0.001363(d + D)^2 L$	$0.00001963(d + D)^2 L$
7. Sorenson	$0.005454(d + L/20)^2 L$	$0.00007854(d + 0.416L)^2 L$
8. Northwest cubic foot log scaling rule	$0.001818[(d' + 0.7)^2 + (d' + 0.7)(D' + 0.7) + (D' + 0.7)^2]$ $(L + 0.67\ \text{or}\ 1.0)$	
9. Hoppus feet*	$(C/4)^2 L/144$	
10. Haakendahl (superficial feet)	$12 * \text{Hoppus}$	

*Gives 78.5% of actual cubic volume of the log.
d = small-end diameter, inches or cm
D = large-end diameter, inches or cm
d_m = mid-length diameter, inches or cm
C = circumference, inches
d' = Scribner scaling diameter small end, west-side method, inches
D' = Scribner scaling diameter large end, west-side method, inches
L = log length, feet or meters

Smalian's formula is one of the most common, sometimes used with the Bruce's variation for butt logs; Smalian itself is quite biased for flaring butts. The subneiloid formula is often incorrectly identified as Smalian's formula. In practice, the subneiloid cubic result is usually multiplied by 12 BF/CF to yield Brererton scale. The Northwest cubic foot rule is an adaptation of the two-end conic formula to permit use of standard west-side Scribner measurements. Hoppus is often used outside North America but fails to meet the criterion that a log rule should measure all of the fiber present, since it incorporates a 21.5% deduction for manufacturing losses.

It is relatively easy to form the ratio of any pair of these cubic formulas or the same one with different measurement systems and get a mathematical result that can be graphed and used as a conversion factor. While often ignored, these differences in cubic measures can be substantial and may amount to 10% or more in certain cases.

Conversion of Scribner to Cubic

The erratic step-function behavior of Scribner, combined with the cylinder assumption, makes the formation of a conversion factor to cubic, or vice versa, very troublesome. Figure 3 illustrates this more clearly for east-side conditions. The points represent individual logs; the great dispersion involved is evident. Cubic volumes were calculated using Smalian's formula. Note the extreme range from a little over 1 to almost 9. Regression equations useful for conversions are also shown in Figure 3. Similar relationships for west-side conditions are presented by Cahill (1984).

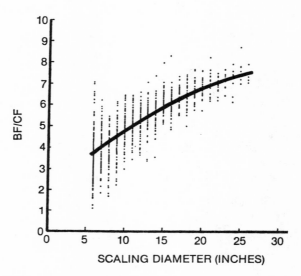

Figure 3. East-side BF/CF ratios plotted over log diameter. Variation in BF/CF is highest for the smaller diameters because of the "step function" characteristics of Scribner scale. This pattern holds true for the other conversion factors reported. Source: Cahill (1984).

East-BF/CF-Gross $= 5.42 + 0.110 \times D - 14.92 \times 1/D$; where $R^2 = 0.70$ and SE $= 0.739$.
East-BF/CF-Net $= 5.33 + 0.085 \times D - 13.93 \times 1/D$; where $R^2 = 0.51$ and SE $= 0.951$.
Log diameter measured using east-side Scribner scaling rules.

West-BF/CF-Gross $=10.52 - 0.0294 \times D - 93.46 \times 1/D + 307.84 \times 1/D^2$; where $R^2 = 0.80$ and SE $= 0.627$.
West-BF/CF-Net $=10.16 - 0.04 \times D - 88.18 \times 1/D + 290.58 \times 1/D^2$; where $R^2 = 0.56$ and SE $= 0.953$.
Log diameter measured using west-side Scribner rules.

An analyst with knowledge of the average log size—or better, the log size distribution—could use these equations to obtain a weighted average conversion factor that would be more realistic than the typical published constant.

An alternative to relying on this approach that would be appropriate to a buyer-seller situation is to agree on a sampling procedure that could be used for each shipment. A fairly small random sample of logs will produce a conversion factor with a specified reliability. Appendix 1 briefly describes the sampling method and Table 3 offers an application of it to a portion of a sample. Here, sample logs were measured by both west-side and east-side methods as well as actual

Table 3. Example of Scribner/cubic conversion with a sample of Douglas-fir logs.

Log	d_L (in.)	d_s (in.)	Length (ft.)	Volume (CF)
			Measurements	
1	20.4	13.8	27.0	45
2	27.5	17.0	41.0	117
3	19.4	12.5	44.3	64
4	22.1	14.3	41.0	77
5	13.0	7.2	20.9	13
6	10.6	6.0	28.4	11
7	27.3	17.7	27.1	78
8	12.8	6.3	23.5	13
9	16.1	10.4	26.9	27
10	23.2	17.4	35.3	81
11	9.0	5.5	39.0	12
12	15.0	7.0	40.9	31
13	17.2	15.0	34.9	50
14	18.3	12.6	34.9	47
15	8.3	6.4	14.9	4
				670

Conversion factor: west-side = 2,810 BF/670 CF = 4.2
east-side = 3,980 BF/670 CF = 5.9

SCRIBNER WEST-SIDE SYSTEM

Scale diameter	Length	Segments	Volume	BF/CF
13	27		160	3.6
17	41	17"×21′, 19"×20′	540	4.6
12	44	12"×22′, 14"×22′	270	4.2
14	41	14"×21′, 16"×20′	350	4.5
7	20		30	2.3
6	28		30	2.7
17	27		310	4.0
6	23		30	2.3
10	26		90	3.3
17	35		400	4.9
5	39		40	3.3
7	40		70	2.2
15	34		300	6.0
12	34		170	3.6
6	14		20	5.0
			2,810	3.8

Table 3. Continued.

SCRIBNER, EAST-SIDE SYSTEM

Scale diameter			Segments	Volume	BF/CF
d_L	d_s	d			
20	14	6	17"×14', 14"×13'	250	5.6
28	17	11	25"×14', 21"×14', 17"×13'	820	7.0
19	12	7	17"×16', 15"×14', 12"×14'	370	5.8
22	14	8	20"×14', 17"×14', 14"×13'	490	6.4
13	7			30	2.3
11	6	5	9"×14', 6"×14'	50	4.5
27	18	9	23"×14', 18"×13'	500	6.4
13	6	7	10"×12', 6"×11'	50	3.8
16	10	6	13"×14', 10"×12'	120	4.4
23	17	6	20"×18', 17"×17'	510	6.3
9	6	3	8"×20', 6"×19'	60	5.0
15	7	8	11"×20', 7"×20'	110	3.5
17	15	2	16"×18', 15"×16'	320	6.4
18	13	5	16"×18', 13"×16'	280	6.0
8	6			20	5.0
				3,980	5.4

dimensions for cubic volume using Smalian's formula. Note the difference between east-side and west-side conversions as well as the difference between the mean factor based on total volumes (correct) and on averaging the factors of the individual logs (incorrect). This sampling procedure has the advantage that variations in species, taper, and so on, are taken into account in each case.

In the foregoing discussion we have not addressed the issue of treatment of scaling defects. West-side and east-side Scribner have different rules for reducing volume for what are called scaling defects (rot, sweep, etc.), and cubic systems may have their own methods. The same procedures described above would be applied, except one must obtain data on defect percentages for the systems and logs in question. Obviously, the sampling approach would be extremely useful in this situation.

Log Weight Estimation

In the absence of direct weighing of truckloads or individual logs as part of a sampling scheme, the conversion difficulties described in the preceding section are compounded by two additional complications in adapting a conversion factor from volume to weight. First, specific gravity or weight density varies greatly within a given species as well as between species. For example, the commonly reported value of specific gravity for Douglas-fir is about 0.45, while the reported range is from 0.36 to 0.54 (Haygreen and Bowyer 1982). Second, wood can vary in moisture content over a wide range. In addition to obtaining appropriate values for these new parameters, one must be aware that each can be measured in different ways (Appendix 2).

Given the specific gravity and moisture content, estimating a weight conversion is relatively straightforward. Figure 3-4 and Table 3-7, in the *Wood Handbook* (Forest Products Laboratory 1974), make this easy for any product. Here, for logs that will always have a moisture content higher than the fiber saturation point, we would base the conversion directly on the SG_{green} and MC_{OD} dry estimates. If the estimates were $SG_{green} = 0.40$ and $MC_{OD} = 80\%$, then Table 3-7 shows 44.9 lb/CF. This is the weight of wood fiber and water present per cubic foot of green or fully swollen volume. To get the weight of 25,000 CF of logs, merely multiply to get 1,122,500 lb. To get the weight per MBF Scribner, divide the weight density by the appropriate BF/CF factor and multiply by 1,000. Using the west-side sample of Table 3 with 4.2 BF/CF, 1,000 BF weighs

$$(44.9 \text{ lb/CF})/(4.2 \text{ BF/CF}) * 1,000 \text{ BF/MBF} = 10,690 \text{ lb/MBF}.$$

The same logs measured by east-side methods give 7,610 lb/MBF.

As an example of equivalent metric calculations, let's start with 708 m^3 of logs. Then the 0.40 SG$_{green}$ can be converted to 400 kg of oven-dry wood per green cubic meter. To this add the water content (i.e., 400 (1 + 80/100)) to get 720 kg of wood and water per green cubic meter. This can be obtained more directly by multiplying the value in Table 3-7 by 16.0256. There may be slight differences due to rounding in the table. Multiplying by the log volume yields 509,760 kg. This same result could also be obtained by remembering that there are 2.2046 lb/kg.

LUMBER CONVERSIONS

In many parts of the world, lumber is manufactured to fairly tight size tolerances and marketed in cubic meters. In these situations a cubic meter of lumber manufactured to a particular degree of surfacing and at a specified moisture content is essentially all wood fiber. Unfortunately, this is not generally true in North America. As a further element of confusion, North Americans measure lumber in board feet. A board foot is supposedly 12 inches square and 1 inch thick, and theoretically there are 12 board feet in a cubic foot. However, with rare exceptions, the use of 12 BF/CF as a conversion factor is invariably wrong. The reasons, for both softwoods and hardwoods, are examined below.

Softwoods

The specifications for softwoods show that actual specified sizes are typically well below the nominal dimensions. Table 4 illustrates this situation for a 2 by 4 cross section. These specifications are minimum sizes; mills will use somewhat larger dimensions (target sizes) at their machine centers to accommodate statistical variability such that only a small percentage is below specifications (undersize). Table 4 indicates that there is a varying amount of wood fiber in a 2 by 4 depending on drying and degree of manufacture. This is best portrayed in the ratio of nominal board feet to actual wood present in cubic feet. These nominal to cubic ratios (NCR) are substantially greater than 12. Note that the units, board feet per cubic foot, have a very different interpretation from the log scale board foot to cubic foot ratio discussed in the preceding section on logs.

Because of the nature of the softwood lumber specifications, the NCR is not constant across all cross sections for a given state of drying and manufacture. Thus shifts in the mix of products will alter the average NCR. The only situation in which the theoretical 12 board feet per cubic foot would be appropriate is in a contract requiring full-size cutting. In this situation actual size equals nominal, a circumstance that exists in some export contracts. Since full-size cutting requires that a mill put more wood in a given cross section compared with traditional specifications, North American mills are often reluctant to lose recovery by taking on such export contracts.

Hardwoods

The specifications for hardwood lumber to be used in the furniture industry generally require the manufacturer to produce a thickness much closer to the nominal size. In addition, since the lumber will be remanufactured into furniture cuttings, widths are random with the convention of rounding to the nearest inch. Thus, for a number of pieces in a given width category, the nominal width and average actual width are essentially identical. Table 4 also illustrates the counterpart cross section to the softwood 2 by 4. In this case, the mean NCR is shown for each condition rather than the range due to random width.

Table 4. Lumber cross section specifications and relation to cubic volume actually present.

SOFTWOOD 2 by 4

	Thickness: 2 inch		Width: 4 inch	
	Dry	Green	Dry	Green
Rough	1.625	1.688	3.625	3.688
Surfaced	1.500	1.562	3.500	3.562

	Board Feet/Cubic Foot NCR
Surfaced dry	18.28
Surfaced green	17.25
Rough dry	16.30
Rough green	15.43

HARDWOOD 2 by 4 (8/4 by 4)

	Thickness: 2 inch (8/4)		Width: 4 inch	
	Dry	Green	Dry	Green
				random
Rough	2.000	2.100		3.5 to 4.5
Surfaced	1.750	1.838		Ave. = 4

	Board Feet/Cubic Foot NCR
Surfaced dry	13.71
Surfaced green	13.06
Rough dry	12.00
Rough green	11.43

*Assumes 5% increase in dimensions to allow for shrinkage. Green size would be specified in a contract arrangement.
Source: Calculations based on NHLA rules and ALS.

The amount of wood fiber that must be present in a piece of hardwood lumber is greater than that in its softwood counterpart, as indicated by the lower NCRs. This means that a softwood manufacturer has a recovery advantage over a hardwood manufacturer in a given log size. It also means that one should not use the same conversion factors for both of these industries.

Relationship to Log Input

Nominal Mill Tally. If logs are measured in a cubic system, the ratio of nominal mill tally of lumber to the cubic log input is called lumber recovery factor (LRF) and is expressed in BF/CF. Note that this is yet another factor with these same units. The reciprocal of LRF indicates the amount of log required to yield a unit of lumber. This gives the roundwood equivalent measure and is one of several reasons why many mills use cubic log measurement for their internal accounting.

If logs are measured in a board foot system, such as Scribner, the ratio of nominal tally of lumber to log scale board feet minus one is the percentage error of the log rule as a model for predicting mill outturn. Many have given this error term, commonly called overrun, much more significance than it deserves. It is highly sensitive to minor changes in log size or product line and its use can be very misleading. If this is the only measure available, try to get the overrun expressed in absolute terms (i.e., board feet) and not in percentage. Overrun in board feet is generally much more consistent across log sizes than is the percentage.

Actual Lumber Fiber. If the actual dimensions of pieces of lumber are measured so their real cubic volume is obtained, the ratio to log cubic volume is the cubic recovery ratio (CRR). This indicates the fraction of the log raw material that became lumber product. This ratio is fairly consistent across product lines and log sizes, although it will vary substantially with degree of manufacture and final moisture content.

In the metric system where actual lumber dimensions are very close to the normal sizes, the ratio of mill tally in cubic meters is the CRR. There is a simple relationship between CRR, LRF, and NCR:

$$CRR = LRF/NCR$$

Table 5 summarizes these lumber factors and when to use them. One must be careful to use any of these factors at the correct level of manufacture and moisture content.

Table 5. Use of lumber volume factors.

Factor	Definition	Typical Range	Use
NCR	Board feet nominal lumber (mill tally) divided by the actual cubic foot volume of the lumber pieces	11-18	•To estimate the actual amount of wood in lumber at a given state of manufacture •To convert nominal lumber volume to actual cubic and vice versa •Use 12 BF/CF only for full-size cutting
LRF	Board feet nominal lumber (mill tally) per cubic foot of log input	6-10	•To estimate lumber output from a given volume of logs •Reciprocal gives log requirement (solid-wood equivalent) to yield a given amount of lumber output
Overrun	Board feet nominal lumber (mill tally) as a percentage of board feet log scale such as Scribner	−10 to 100+%	•The error of the log scale as a lumber model •Very sensitive to slight changes in product line and log size
CRR	Cubic feet of actual wood in lumber per cubic foot of log input	0.4-0.7	•To estimate % of log processed into lumber versus residues

Note: NCR and CRR may be calculated for the following states: rough green, surfaced green, rough dry, surfaced dry. The BF/CF ratios should not be confused with the log rule BF/CF ratio used to convert Scribner log volume to cubic log volume.

Since the factors are sensitive to product line, degree of manufacture, moisture content, and log size, one must be careful in specifying objectives and conditions. A frequency table of cross sections is especially useful in getting a good weighted average, and sampling may be useful in gathering some data.

Lumber Weight Estimation

Considerations for lumber weight are similar to those for logs except that one often has better information on moisture content if the lumber has been dried. In obtaining a weight conversion, one must be careful to note that below the fiber saturation point wood begins to shrink. In doing so, its volume changes and one must adjust for this.

As an example, suppose that one has 1,000 board feet of surfaced-dry 2 by 4s that have been dried to 15% MC_{OD}. Further, suppose that the SG_{green} for the species is 0.40. Figure 3-4 and Table 3-7 in the *Wood Handbook* can be used with this information to develop the appropriate weight. First, adjust SG_{green} to $SG_{15\%}$ by following along the SG = 0.40 line of Figure 3-4 to 15% MC_{OD} where one reads $SG_{15\%}$ = 0.42 from the left axis. Next, in Table 3-7, find the intersection of SG = 0.42 and MC_{OD} = 15 and read about 30.2 lb/CF. This is the weight of wood fiber and moisture per cubic foot of wood volume as measured at 15% moisture content dimensions. Finally, in Table 4 (above), read that surfaced-dry 2 by 4s have an NCR of 18.28 BF/CF. Dividing, we find that the 1,000 BF contains 54.7 CF of solid wood. The actual weight of 1,000 BF of surfaced-dry 2 by 4s of the particular species is 1,652 lb.

Suppose one already had converted the solid lumber volume to 1.55 m^3. The wood density in metric units is 30.2 * 16.0256 = 483 kg. Therefore, the weight of the lumber is about 748 kg. Alternatively, one would get the same result using the 2.2046 lb/kg.

CHIP AND RESIDUE CONVERSIONS

Chips and residues are sold either in terms of volume measures, such as the 200 cubic foot unit (5.66 m^3) or cubic meter (0.117 unit), or in terms of weight, generally the bone-dry unit, BDU (2,400 lb, 1,088.62 kg) or the bone-dry tonne, BDT (1,000 kg, 2,204.623 lb). Since these latter represent the weight of oven-dry (0% MC) wood, the shipping weight is considerably higher because of moisture.

The potential for confusion in working with chips and residues is great because of the lack of precise definition of terms in many reports plus the desire at times to express these conventional measures in short tons. In addition to the uncertainty as to the definitions of reported values, development of a conversion factor usually requires data or assumptions regarding moisture content, wood specific gravity, and bulk density (Appendix 2). Bulk density increases with increased compaction and is influenced by specific gravity, moisture content, and particle geometry.

Conversions are typically desired between the standard chip volume and weight measures as well as solid-wood equivalent, short tons, long tons, and so forth. The procedures can be summarized in either of the following equations:

$$(1) \quad W = C * B * k$$
$$(2) \quad W = S * D * k$$

where W = weight, lb or kg

C = particle volume, CF_p or CM_p

B = bulk density, lb/CF_p or kg/CM_p

S = solid-wood equivalent, CF_s or CM_s

  = C/F

F = bulk density, CF_p/CF_s or CM_p/CM_s

D = wood density, lb/CF_s or kg/CM_s

k = 1 if units are lb and CF

  = 16.0256 if units are kg and CM

One can substitute more detailed expressions for D involving specific gravity and moisture content. Also, if one substitutes for F, it appears that various terms may cancel. But this is valid only if W is desired under exactly the same moisture content conditions that B or F was measured in. It should be apparent that one can easily rearrange terms in the equation to find out, for example, how many units are represented by a certain weight of chips. Table 6 presents an example illustrating the calculations.

Table 6. Chip conversion example.

Data

      15 units chips = 3,000 CF_p = 85 CM_p
      bulk density = 28 lb/CF_p = 448 kg/CM_p
      SG_g = 0.5
      Table 3-7 (*Wood Handbook*) Wood density
              0% MC   31.2 lb/CF_s = 500 kg/CM_s
              80% MC   56.2 lb/CF_s = 900 kg/CM_s

Equation (1) W = C * B * k based on particle volume
      shipping weight   = 3,000 * 28 * 1 = 8,400 lb
                                   = 85.0 * 28 * 16.0256 = 38,140 kg
      oven-dry weight   = shipping weight/(1 + MC_{OD}/100)
                                   8,400/1 + 80/100) = 46,667 lb or 19.4 BDU
                                   38,140/(1 + 80/100) = 21,189 lb or 21.2 BDT

Equation (2) W = S * D * k based on equivalent of particles
      F = WD/BD = 56.2/28 = 2.00 CF_p/CF_s
      S = 3,000/2.00 = 1,500 CF_s or 42.5 CM_s
      oven-dry weight   = 1,500 * 31.2 * 1 = 46,800 lb or 19.5 BDU
                                   = 42.5 * 31.2 * 16.0256 = 21,249 kg or 21.2 BDT
      shipping weight   = 1,500 * 56.2 * 1 = 84,300 lb
                                   = 42.5 * 56.2 * 16.0256 = 38,280 kg

---

Note: Values obtained by the different methods may vary due to rounding inherent in Table 3-7.

SUMMARY AND CONCLUSIONS

Having examined the conversion factors for these examples, it's apparent that each involves a number of critical variables and as a consequence it would be rare for a single conversion to apply universally. It is ironic that in today's modern high tech world the forest industry takes such pains to squeeze processes down to the last few thousandths in order to improve recovery, product uniformity, and performance and then relies on archaic, crude, and variable systems of measurement in commerce. Perhaps adherence to these systems is some sort of rite-of-initation process to become a member of the wood products club, but it is frightening to see the level of ignorance within and outside the industry as to how measurement systems actually operate. While a simpler standardized global system would be welcomed by many, it is doubtful that changes will occur soon. There are too many vested interests and perceived advantages of the current situation. Examples are (1) the ability to hide real quantities from tariff and tax agencies by choosing the most advantageous combination of measurement systems, (2) the marketing advantage one may have over others, particularly newcomers (countries may also use their systems as a way to create an artificial trade barrier by haggling over measurement definitions), and (3) the ability to fleece unsuspecting customers or resource owners.

These aspects may cause more problems than they are worth by placing an additional degree of uncertainty on buyer-seller transactions. They may place forest products at a disadvantage compared with other materials if architects, engineers, traders, and others find that forest products are too difficult to learn and understand.

The purpose of this paper has not been to propose any new solutions to conversion factor problems but to indicate how better conversions can be derived and some of the variables involved. It is hoped that this discussion will bring about more rigorous thinking on the conversion needed in a given situation and lead to refined estimates. One can often devise simple spread-sheet programs to develop good factors and quickly update them. The authors suggest using sampling procedures as a useful way to develop and monitor conversion factors, and would like to recommend the following points:

1. Ask many questions to gain understanding of measurement methods, calculation basis, and final units. For example, there are various interpretations of the BF/CF ratio as well as lb/CF.

2. Ask how much conversion error can be tolerated and who absorbs the cost.

3. In many cases organizations may already collect useful information but communication may be poor. For example, quality control sampling may already gather the needed data or it may be easy to append collection of additional data onto quality control. It may take less time and yield better results to sample the shipments rather than rely on published reports. Use such published factors only as a last resort.

4. Those who publish should be encouraged to be more responsible to document whether the information presented is in its original or converted form and, if information has been converted, the basis for the conversion.

5. Some conversion factors will change over time because of changes in the resource being used. Sample monitoring would be a valuable basis for tracking and projecting such changes.

REFERENCES

Avery, T. E. 1975. Natural resources measurements. McGraw-Hill, New York.

Cahill, J. M. 1984. Log scale conversion factors. *In* T. Snellgrove et al. (eds.) User's guide for cubic measurement. Contribution 52. College of Forest Resources, University of Washington.

Darr, D. 1984. Conversion factors can affect forest products trade policies. J. For. 82(8):489-491.

Durst, P. B., C. D. Ingram, and J. G. Laarman. 1986. Statistics on forest products trade: Are they believable? *In* G. F. Schreuder (ed.) World trade in forest products 2, pp. 265-273. University of Washington Press, Seattle.

Forest Products Laboratory. 1974. Wood handbook. Agric. Handbook 72. USDA Forest Service, Forest Products Laboratory, Madison, Wisconsin.

Hartman, D. A., W. A. Atkinson, B. S. Bryant, and R. O. Woodfin. 1976. Conversion factors for the Pacific Northwest forest industry. College of Forest Resources, University of Washington, Seattle.

Haygreen, J. G., and J. L. Bowyer. 1982. Forest products and wood science. Iowa State University Press, Ames.

Koch, P. 1985. Utilization of hardwoods growing on southern pine sites. Agric. Handbook 605. USDA Forest Service.

Statistics Canada. 1977-85. Exports by commodities. Statistics Canada, International Trade Division, Ottawa. December issues.

Warren, D. D. 1987. Production, prices, employment and trade in Northwest forest industries, third quarter 1986. Resource Bulletin 142. USDA Forest Service, Pacific Northwest Research Station.

APPENDIX 1

A Sampling Method to Obtain a Conversion Factor

The formulas for simple random sampling are (Avery 1975):

(1) finite population $\quad\quad n = (Nt^2C^2)/(NA^2 + t^2C^2)$

(2) infinite population $\quad\quad n = (tC/A)^2$

where

   n   = number of random samples to take

   N = population size. Use infinite population formula if N is large.

   A = acceptable error that can be tolerated in the conversion factor. This is the standard error of the mean expressed as a percentage of the mean.

   C = coefficient of variation of the conversion factor,
      = standard deviation/mean * 100. This can be estimated from a small preliminary random sample or from past experience.

   t   = Student's t statistic for a given level of probability, alpha

Example: N very large, A = 2%, C = 15%, alpha = 5%

$$n = (1.96 \times 15/2)^2 = 216$$

Suppose the mean conversion factor of the sample turns out to be 5.9 BF/CF, then one can be 95% certain that the sample mean is within 2% of the true conversion factor of the population.

If you vary A while holding C at 15% and alpha = 5%, you get

Allowable Error %	n
2	216
3	96
4	54
5	35

Density: weight per unit volume

Specific gravity: density of wood/density of H_2O
    Conventions:
            Always use oven-dry weight in the wood density part.
            Must specify a moisture condition since wood shrinks/swells below fiber
            saturation point (~30% MC_{OD}).
Example: oven-dry sample weighs 75 g
(1) volume fully swollen (green) = 150 cc
    SG_{green} (75 g/150 cc)/(1 g/cc) = 0.50
(2) volume oven-dried to 15% MC_{OD} = 142 cc
    $SG_{15\%}$ = (75 g/142 cc)/(1 g/cc) = 0.53
(3) volume when dried to 0% MC = 135 cc
    $SG_{0\%}$ = (75 g/135 cc)/(1 g/cc) = 0.56
Note: SG increases as MC_{OD} decreases below fiber saturation point and is constant
    (i.e., SG_{green}) above fsp
Formula to convert between SG bases

$$SG_x = SG_{green}/\left(1 - \frac{\% \text{ vol. shrink from fsp to x}}{100}\right)$$

Alternately use Figure 3-4 of the *Wood Handbook* to estimate these changes.

Bulk density:
(1) weight of particles per unit of volume occupied
Example: 20-30 lb/CF of particle occupied space (Koch 1985)
(2) volume of particles divided by volume of equivalent solid wood (Hartman et al. 1976)
Example: 2-3 CF particles per CF solid wood
    These values can be obtained experimentally by taking a known volume of solid wood
    and measuring the space occupied after fragmentation into particles. They also can
    be derived by dividing the bulk density in part 1 by the corresponding wood density
    from Figure 3-4 and Table 3-7 of the *Wood Handbook* for the appropriate particle
    moisture content.

Moisture content:
(1) oven-dry weight basis—usually solid-wood products industries
    $MC\%_{OD}$ = (weight of H_2O/weight of oven-dried wood) × 100
(2) total weight basis—usually pulp and paper, chips and residues
    $MC\%_{TOT}$ = (weight of H_2O/weight of oven-dried wood + weight of H_2O) × 100
(3) Example: original weight   =   125 cc
                  oven-dry weight   =   <u>75 cc</u>
                  H_2O weight   =   50 cc

                  MC_{OD} = (50/75) × 100   = 67%
                  MC_{TOT} = (50/125) × 100 = 40%
(4) Conversion: $MC\%_{OD}$ = $MC\%_{TOT}/(100 - MC\%_{TOT})$ × 100
                  $MC\%_{TOT}$ = $MC\%_{OD}/(100 + MC\%_{OD})$ × 100

Chapter 13

Improved Lumber Trade
through Better Grading Rules

BORG MADSEN

Grading rules constitute a very important element in the trade of lumber, because they succinctly define the products the sawmill has to deliver to the customer. But that is so obvious that most people take it for granted. Yet this is an area in which the sawmiller can improve the profitability of his operation and place his customer in a better competitive position in relation to other materials.

Lumber, unlike other structural materials, is obtained by merely cutting up the logs into smaller pieces. With manufactured materials such as steel and concrete it is possible to change the strength properties by altering the mix of compounds, or some other parameter such as temperature, in the manufacturing process. This cannot be done with lumber, and the only way one can enhance the strength properties of lumber is to sort the desirable pieces by grade and reject the others. In doing so the cost of the rejected material must be compensated for by a premium price for the selected material.

The grading rule is—or should be—a compromise between what the customer needs and what can be produced economically by the sawmill. This is a delicate balance which should be fine-tuned occasionally, since both the supply of logs and the needs of the customer may change with time.

Many different grading rules have been developed, in fact at least one for each major use of lumber product (railway ties, ladder stock, construction lumber, etc.).

This paper will deal with material intended for structural purposes, which covers roughly three-quarters of the overall sawmill production. Discussion will be limited to softwoods (conifers). While some of the generalities of grading may also apply to hardwoods, the specifics may not apply.

STRUCTURAL GRADING RULES

Structural grading rules deal with two types of requirements: those related to strength and those related to appearance or shape. The natural growth characteristics affecting strength include knot size, knot location, general slope of grain, localized slope of grain, and presence of splits, among others. The appearance or cosmetic requirements are wane, warp (bow, crook, twist), smoothness of surface, machining defects, and so forth.

The strength requirements in the grading rules are reflected primarily by knot size, even though recent research has shown that size by itself is not that reliable a strength indicator. What is more important is the steepness of grain disturbance associated with the knot.

For structural applications it is imperative to have correct strength values. If they are set lower than what they could be, the lumber is used in a wasteful manner that reduces its market potential.

If the strength properties are set too high, the user and the producer may face increased liability in case of a failure.

In this connection it should be emphasized that drastic changes are taking place in the engineering design methods, with much greater attention to reliability. That in turn makes it important that representative and reliable strength information be made available to the design engineers.

NORTH AMERICAN GRADING RULES

The grading system used in North America is based on the strength of *clear* wood, which varies greatly from species to species. The strength of small clear wood specimens is determined through tests, and the fifth percentile of the strength values is determined as representative of a minimum strength. To get to the allowable stress for a specific grade, the clear wood strength is multiplied by a strength ratio that reduces the strength further to account for maximum knot size allowed in the grade.

Four structural grades are defined in the grading rules, and some information is given in Table 1 pertaining to those grades. In the second column, the strength ratio for the grade is shown. Values for allowable stress are shown for two species, Douglas-fir and S-P-F (spruce-pine-fir), in the next columns. The same data are shown in graphic form in Figure 1. The allowable stresses based on small clear tests for Douglas-fir grades are stronger than for the S-P-F grades, category by category. The categories (i.e., Select Structural, No. 1, No. 2, No. 3) decrease monotonically in the old system (small clear basis).

Table 1. Grading rules for four structural grades of Douglas-fir and spruce-pine-fir.

| Grade | Strength Ratio (%) | Allowable Stress (psi) | |
		Douglas-fir	S-P-F
Select Structural	65	1,850	1,300
No. 1	55	1,600	1,100
No. 2	45	1,300	900
No. 3	26	750	500

The in-grade testing program conducted in Canada ten years ago using the end product (full size lumber) gave quite different results from those described above. The allowable strengths based on the in-grade results for Douglas-fir and S-P-F, 2 by 8, are also shown in Figure 1. Only a small difference between the two species groups can be seen. Moreover, the strength difference between the grade categories is not consistent.

Obviously the present grading system (small clear basis) does not enable us to utilize the lumber to its fullest extent. As a result, lumber is less competitive in relation to other materials and we are not picking up the profit that could be available to the lumber industry.

But what can we do about it? Technically it is a relatively simple task to develop improved visual grading rules, but it may be more difficult to overcome the inertia of traditional thinking.

PILOT STUDY OF VISUAL GRADING RULES

A project was carried out recently at the University of British Columbia aimed at improving the visual grading rules used for structural lumber (Madsen 1985).

The project commenced with a review of the needs in the marketplace. The analysis showed that there were essentially three different end uses: (1) engineering uses (15-20%), which include

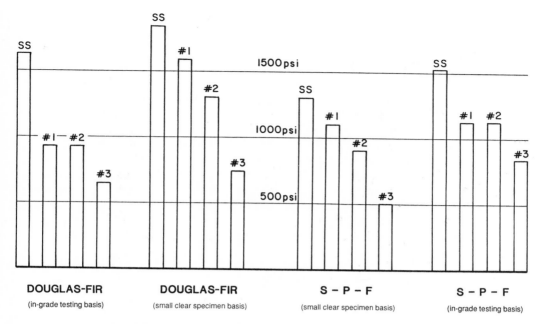

Figure 1. Comparison of strengths for Douglas-fir and S-P-F.

trusses, wood I-beams, and engineered buildings; (2) housing uses (60-70%), which include single and multiple family housing and apartments; (3) general purpose uses (15-25%), which include home improvements built by the owner. Each category was then analyzed with regard to structural and cosmetic requirements. The requirements were quantified to minimize ambiguity.

The second step was to obtain 800 pieces of 2 by 6 Hem-Fir KD from a typical mill run. It was essential for the pieces to be obtained before any grading had taken place in order for the subsequent yield calculations to be representative. The mill personnel were then asked to remove the pieces that would fetch a premium price, such as "clears." Now 720 pieces were left, representing that portion of the mill run available for the three identified end use categories.

The third step was to apply all available grading systems to the 720 pieces. Australian rules, New Zealand rules, European rules, and so forth, were applied together with North American rules. The grading was done both with and without cosmetic requirements, and the pieces were run through two mechanical grading machines, the CLT-1 and the Cook-Bolinder machine.

The reason for all this grading was to compare the effectiveness of the existing grading rules and identify the best ones.

The pieces were randomly divided into two groups: group A and group B. The pieces in group A were tested to destruction using a standard bending test with a span-to-depth ratio of seventeen to one. The broken pieces were laid out side by side in the yard in increasing order of strength so that the cause of failure was readily apparent for each piece.

The analysis of the end use requirements had shown that maybe 20% of the pieces of the mill run could be sold for engineering applications; because strength is of primary importance, the strongest 20% of the pieces were marked off by placing a marker 20% down from the top. Since the market analysis also showed that the Housing grade should have a strength of at least 1,100 psi, a second marker was placed between the pieces having a strength of 1,100 psi.

Four teams were then asked to develop visual grading rules that would segregate the top 20% of the pieces from the rest and similarly develop rules that would segregate pieces weaker than 1,100 psi from the stronger ones. Two of the teams consisted of experienced lumber graders and two teams of students with no previous grading experience. The four resulting grading rules were then applied to the pieces in group B and the pieces broken to see how well the new grading rules performed.

Based on the existing grading rule and the best of the four newly developed rules, we could calculate the yield and the characteristic strength (fifth percentile) associated with each grade. The strength was assigned a monetary value which, together with the yield, enabled us to calculate the value of the mill run for each grading rule.

The comparison showed that there was not much difference between the existing visual grading rules but that the best of the four new grading rules provided a substantial improvement.

Table 2 gives a summary of the results. Using No. 2 and Better with stresses according to CSA 086-1980, the mill run had a value of $143.20. The introduction of new CSA strength properties obtained from the in-grade testing changed that value to $154.70. If the actual fifth percentile had been used rather that a conservative estimate of the fifth percentile, the value of the mill run would have been $180.40. Using the new grading rules the value increased to $206.10. It should be noted that the mill run graded by a mechanical grading machine would fetch a value of $178.70.

Table 2. Value of mill runs according to degree of grading.

Degree of Grading	Value of Mill Run	
	Hem-Fir	S-P-F
No. 2 and Better, No. 3, Rejects CSA 086-1980 stresses	$143.20	$161.40
No. 2 and Better, No. 3, Rejects CSA 086-1984 stresses	$154.70	$182.10
No. 2 and Better, No. 3, Rejects (fifth percentiles)	$180.40	$195.20
Engineering, Housing, General Purpose (fifth percentiles)	$206.10	$221.80
M.S.R. lumber (machine stress rating)	$178.70	$213.20

The procedure was repeated with a similar sample of 2 by 4s of S-P-F, and in both cases the value of the mill runs increased by 15% merely by updating the usual grading rules.

ISO GRADING SYSTEM

Technical Committee 165 of the International Standards Organization is working on a new grading system that could become very useful for countries exporting or importing lumber. It differs from the present visual grading rules in that it is a performance standard rather than a descriptive standard. It simply defines some strength classes together with test methods to verify that the material meets the strength requirements. It is then up to the sawmill to develop a method for selecting the desired pieces which fits that particular mill. The system is independent of species and sizes and gives the sawmill the freedom to use the strength classes most suitable for its production. From a design point of view it is a desirable system because it provides opportunities for easy substitution if availability is a problem.

The system separates building regulations from the grading system, and that is of considerable advantage for the code authorities as well as for the grading rule agencies. (A change in one does not affect the other, since the common denominator is the strength class and that remains unchanged.)

The strength classes are arranged as a geometric series, which is mostly effective for structural design (each strength class is 25% stronger than the previous one). Even though some details remain to be worked out, the basic idea has already been incorporated into the CIB W18 Timber Design Code. Figure 2 shows the strength classes and the characteristic strength for the mill run described earlier when graded according to the new visual grading rules.

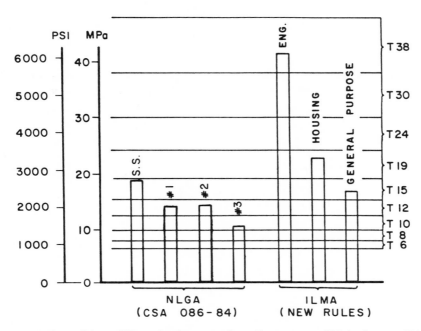

Figure 2. A comparison of three different lumber strength grading systems: ISO is shown as T6 through T38, ILMA (Interior Lumbermen's Association) is shown as Engineering, Housing, and General Purpose, and NLGA (National Lumber Grades Authority) is shown as SS, #1, #2, and #3.

RELIABILITY

It was mentioned earlier that reliability is becoming a much more important aspect of structural engineering design. For timber it will result in improvements to some of our design methods so they will more accurately reflect the structural behavior of lumber or timber. Over the next three to seven years new formulas for wood columns will be introduced and length effects will be included as a design consideration for bending and tension. The factors for duration of load will be updated and so will the adjustment factors for moisture content. In fact, the whole gamut of timber engineering is being examined for improvements. It would be very appropriate if the grading methods could also be subjected to a constructive scrutiny so that the lumber industry could have the benefit of a thorough updating. Without it, it may be more difficult to convince structural engineers that lumber products do present an attractive alternative to other structural materials.

SUMMARY

The purpose of this presentation was to draw attention to the tantalizing prospect of improved profits by updating the visual grading rules. Admittedly it is a large task because so many people are involved, but technically it is both desirable and feasible. It would, at the same time, make timber a more reliable structural product and therefore more attractive to the structural designer.

The pilot study shows that it can be done. What we in the engineering and technical side are looking for is a positive response from those involved in the marketing aspects so that both groups can work toward substantial increases in the profitability of the manufacturing and trading of structural lumber.

REFERENCE

Madsen, B. 1985. Pilot study for evaluating and developing visual grading rules for lumber. Department of Civil Engineering, University of British Columbia, Vancouver.

Chapter 14

National Codes and Standards: Help or Hindrance in the Utilization of Wood Products for Construction Uses

C. K. A. STIEDA

The development of standards is intimately connected with two basic human activities, building and trading. As the Roman architect Vitruvius observed about two thousand years ago, the human need for protection from the hostile forces of the environment has led to the development of measurements for construction of shelter based on human proportions: the width of a digit, the length of the foot or that of the forearm, a single stride or a double pace, the distance between the tips of the middle fingers of the outstretched arms. It is likely that the person who needed a house would ask the builder not just to build him a house but a house twice as big as that one over there, so many paces long, and so forth. From the need to agree on such measurements, standards were born.

One of the remarkable features in the development of units of length is how similar these units are that were developed by different people in different areas of the world. The Egyptian cubit, equal to the length of the human forearm, has its equivalent in the German *Elle*; the Anglo-Saxon foot is nearly identical to the Japanese *shaku*.

Concern for just measurements—standards in today's language—and fair trade practices are well recorded in old documents. In Leviticus we read the following exhortation: "You shall do no wrong in judgment in measures of length or weight or quantity. You shall have just balances, just weights, a just ephah and a just hin" (Lev. 19:35). Of course, different standards, based on local customs, developed for each of these units. And this must have resulted, as we can well imagine, in numerous trade disputes.

This paper will look at a number of examples of local standards for length measurements and the growing need to use the standards developed for the measurement of force, mass, and length in the International System of Units, the so-called metric SI units. The description of forest products traded nationally and internationally usually is done in terms of quantities that can be measured, be it the volume of lumber, the weight of pulp, or the bond strength of plywood. Some examples will be given of existing differences in national product standards, their origins, and their effects on international trade. Finally, the utilization of wood products in construction will be considered, examples of national design codes will be given, and the efforts to harmonize national design codes will be described.

BASIC UNITS OF MEASUREMENTS

Early units of measurement were related to the dimensions and proportions of the human body. The usual length of the human stride is three feet, hence the subdivision of one yard into three feet. The width of the human palm is about one-third the length of a foot. The four fingers attached to the palm suggest immediately a subdivision into four units each equal to the width of a digit. The width of a digit therefore is equal to one-twelfth the length of the foot. This width has been called an inch, a word derived from the Latin *uncia*, meaning one-twelfth of something. This process of using proportions found on the human body therefore has lead to the subdivision of the foot into twelve inches.

Twice the length of the foot is the usual length of the human forearm, a unit often called a cubit. Of course not all forearms are the same length. And so we find reference standards were developed, such as the standard of Amenophis I (1546-1526 B.C.) for the royal Egyptian cubit. This reference standard was made of wood and is now kept at the Louvre in Paris (Skinner 1967). Many slightly different cubits are recognized by historians, such as the Roman, Egyptian, Greek, Assyrian, Sumerian, Talmudic, and Palestinian cubit. The royal Egyptian cubit was approximately 500 mm long, and over its more than three-thousand-year history in different cultures has not varied more than ±5% (Klein 1975).

According to Skinner (1967), the cubit was also used in ancient China. Engel (1964) indicates that the *shaku*, the name for the Japanese unit of length equivalent to a foot, was introduced from China. And since one cubit equals the length of two feet, it might be speculated that indeed the concept of using the lengths of the human foot and forearm as units of length found their way from ancient Egypt via China to Japan.

Two other units of measurement unique to Japanese house construction should be mentioned here, the *ken* and the *jo*. Both are discussed in detail by Engel (1964). The ken is a measure of length and usually designates the center-to-center distance between two adjacent wooden posts in the wall of a house. The ken is subdivided into 6 shaku of 303 mm length each. One ken therefore equals 1,818 mm. As we will see later, this length still appears in today's standards for plywood in Japan.

The jo is the area covered by one tatami mat. This concept of a standard area is unique to Japanese building construction. In fact, there are numerous local variations in size of the tatami mat. But the mat used in the so-called *inaka-ma* method of construction measuring 3 by 6 shaku is usually recognized as the standard size for a mat.

According to Engel (1964), the jo, the size of a tatami mat over the centuries, has had a profound influence on Japanese architecture. Even today, room sizes are usually expressed as multiples of a mat. A reading of Vitruvius shows the concern Greek and Roman builders had for achieving harmonious proportions in their constructions. Their quest for a standard of length, a modulus, is mirrored by today's efforts of architects to find a suitable modular dimension that will meet the needs of industrialized construction in the twentieth century. The modular unit of a tatami mat as used in Japan perhaps comes closest to this quest for a standard that reflects the human scale of measurements. It is probably not without coincidence that this modular unit, the tatami mat, is not an absolute standard, but in its various forms allows for the large variety of human needs and wishes.

As has been outlined, units for the measurement of length developed by different cultures, while surprisingly similar in concept, have not been standardized against each other. To overcome the many obstacles to trade inherent in the use of measuring systems different from country to country, a metric system of measurement was proposed and adopted in France six years after the French Revolution of 1789. To provide a truly international standard it was proposed to use the distance from the earth's pole to the equator as a reference standard and to make the basic unit of length, a meter, equal to the fraction of one in ten million of this distance. A detailed account of this development can be found in an appendix to ASTM Standard E 380 (1984).

Coupled with the development of this new standard of measurement was the idea of using a decimal system rather than the duodecimal measurement system of feet and inches. The idea of a decimal system was not a new one. Certainly the ten digits of human hands must have suggested counting in groups of ten many times before. James Watt suggested such a system in 1783. An electronic pocket calculator is ample proof that the decimal system has become the universal standard for our system of counting.

Today's internationally agreed system of measurements is known as the SI system (Le Système International d'Unités). This system provides for seven base units, including the meter as a measure of length and the kilogram as a measure of mass, from which all other units of measurement are derived. Of the derived units the most important one for structural engineering is the unit of force, newton (N). Strength values for wood, for example, are expressed using the term pascal (Pa), which is a unit of pressure equal to one newton per square meter. One million pascals (1 MN/m^2) is the equivalent of 145 pounds per square inch. Canadian building and design standards have published strength values for wood in SI units since 1980. No other units are being recognized by building codes in Canada. All countries importing wood products from North America use SI metric units. Adherence by the United States to the inch-pound system of measurements amounts to a self-imposed barrier to lumber exports. However, while removal of this particular barrier to a free flow of wood products across national borders would seem to be a step in the right direction, it is not, as will be shown below, the only existing trade barrier presented by standards which is impeding international trade in wood products.

STANDARDS FOR MEASURING THE GLUE BOND OF PLYWOOD

To buy a pig in a poke is not considered a wise action. The builder purchasing some plywood for the construction of a house would like to be assured that this material will perform as well as the plywood he bought a year ago for the construction of another house, and the structural engineer who designs a stress-skin panel would like to be sure that the material that will be used by the fabricator will indeed have the strength implied by the building code. On the other hand, the manufacturer who has to compete with the suppliers of other building products would like to tell potential users exactly what his product is. And if the manufacturer plans to stay in business, he wants to make sure that the product he supplies today is at least as good as the one he supplied a year ago.

To meet these various objectives, suppliers and users agree on a product standard. In the case of plywood such a standard might give information on the panel dimensions, the available thicknesses, the quality of the plywood surfaces, the wood species used in its manufacture, and the quality of the bond between individual plies of wood. Such product standards usually are developed by independent bodies, such as the Canadian Standards Association or the National Bureau of Standards in the United States, on the advice and with the full cooperation of both users and producers. Since similar organizations are found in most industrial countries, national standards for plywood exist for many countries, such as the ones listed in Table 1.

These national standards provide numerous detailed requirements that describe what are considered to be characteristics that must be met if the plywood is to be considered standard plywood. One of the most important of these characteristics is the bond strength of plywood, the ability of the adhesive used in its manufacture to hold the individual veneers together. To ensure that plywood is manufactured with a durable bond between the veneers, certain requirements for the bond are specified in the various national plywood standards. Table 2 lists the requirements of five countries. A comparison of the these requirements will show two different philosophies in the assessment of bonds. All five standards require some treatment of test samples, usually involving alternate wetting and drying, sometimes including a cycle in which the sample is immersed in boiling water.

Typical bond specimens are shown in Figure 1. The specimen itself is roughly 25 mm wide and 75 to 100 mm long. As can be seen in Figure 2, there is no agreement on the exact sizes of this test specimen. At least two grooves (in the case of the French specimen, three grooves) are required to prepare the specimen for testing. The test itself is illustrated in Figure 3. A tensile force applied to the ends of the specimen will separate the specimen into two halves roughly along the contact area where two adjacent veneers are bonded together.

Table 1. Types of construction plywood covered by standards in Canada, the United States, Germany, and Japan.

Standard	Title and Type of Plywood
CSA Standard 0121 CSA Standard 0151	DOUGLAS-FIR PLYWOOD CANADIAN SOFTWOOD PLYWOOD Species: 18 coniferous, 2 deciduous Adhesives: moisture/temperature resistant

U.S. Product Standard PS 1-83	CONSTRUCTION AND INDUSTRIAL PLYWOOD
	Species groups: 5
	Species: 70
	Plywood type/adhesive
	      Interior/interior
	      Interior/intermediate
	      Interior/exterior
	      Exterior/exterior

DIN 68 705, Part 3	SPERRHOLZ, BAU-FURNIERSPERHOLZ BFU 20 nonweather-resistant bond BFU 100 weather-resistant bond BFU 100G decay-resistant species or preservative treated
JAS 894	STANDARD FOR STRUCTURAL PLYWOOD Class 1 Class 2 Grades 8 3 Face veneer grades: differ Bond requirements: same Strength requirements: differ Species: determine bond strength

Table 2. Bond requirements for structural plywood in Canada, the United States, Germany, Japan, and France.

| | | | | Percentage of Panels in Lot Required to Meet Wood Failure | | | | | | | | | |
| | | | Bond | | | | | % Wood Failure | | | | | |
Standard	Type of Panel	Type of Adhesive	Strength (N/mm^2)	25	30	34	45	50	60	65	75	80	85
CSA 0121	--	Exterior			95				90			M	
U.S. PS 1	Interior	Interior			90		M						
	Interior	Exterior			95				90			M	
	Exterior				95				90			75	M
DIN 68 705	density	<0.5 g/cm^3	1.0										
		>0.5	0.8										
JAS 894 (new)			0.6				X						
			0.5							X			
			0.4									X	
Placos 894			2.5	X									
			2.0		X								
			1.5			X							

M = mean of all specimens   X = permitted minimum strength/% wood failure combination.

The interpretations of the results of this bond test are quite different. In North America the concern is to ensure that the glue bond is at least as strong as the wood itself. We therefore look at the percentage of wood failure on the failure surface and interpret high percentage of wood failure as a "good" bond. Given the rough treatment that the specimen receives as a result of the wetting and drying cycles, it is reasonable to assume that a high percentage of wood failure implies a low probability of adhesive failure during the service life of the plywood. Long time exposures of plywood to changing environmental conditions seem to have confirmed this interpretation.

In Canada only one type of adhesive is recognized by the standard, a so-called exterior type. Plywood manufactured with such an adhesive can be used both indoors and outdoors, and a high average wood failure for the bond test is required.

In addition to the exterior type of adhesive, the U. S. product standard PS 1 also recognizes an interior type of glue that is less durable under exposure to moisture. In combination with certain requirements for the veneers, this has resulted in three different types of plywoods that can be manufactured for construction purposes in the United States. The bond requirements for these three

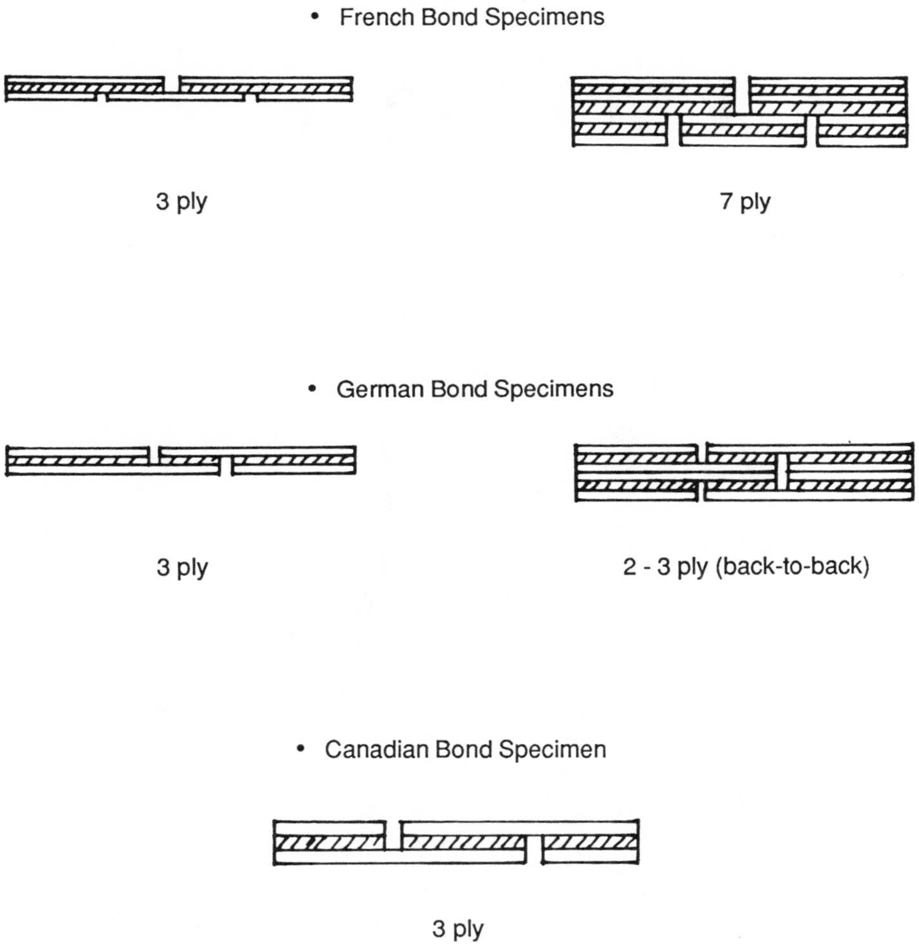

- French Bond Specimens

3 ply   7 ply

- German Bond Specimens

3 ply   2 - 3 ply (back-to-back)

- Canadian Bond Specimen

3 ply

Figure 1. Various plywood bond specimens.

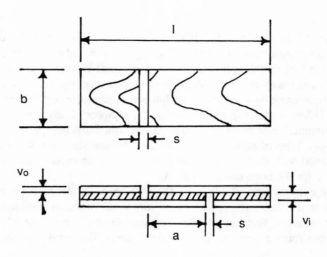

Country	Standard	Units	Veneer Thickness		Length	Width	Cut	Shear Width	Plies
			face	inner					
			v_0	v_i	l	b	s	a	
Canada	CSA 0121	mm			81	25	3	25	multiple
U.S.	PS1 - 83	in			3.25	1.0	.125	1.0	multiple
		(mm)			(83)	(25)	(3)	(25)	
Germany	DIN 53 255	mm		>0.8	100	25	3	10	3
Japan	JAS 894	mm	>1.6		75+	25	--	25	3
France	NFB 51 - 338	mm			150	19.5	4.5	20	multiple

Figure 2. Dimensions of plywood bond specimens.

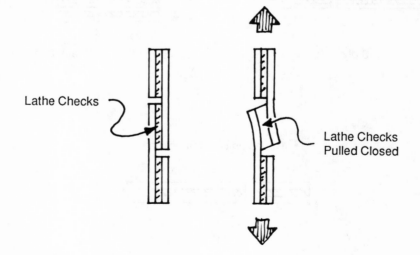

Figure 3. Plywood bond test.

types of plywood vary from a low percentage of wood failure for the mean from all test specimens of interior type panels with interior adhesive to a high percentage of wood failure for exterior panels.

Such classification of plywood bonds on the basis of percentage of wood failure is not recognized in Germany, where the strength of the plywood bond is the characteristic being measured. This requirement originated with the wide use of beech veneers for the manufacture of plywood in Germany. Beech, a very strong hardwood, will not fail in the wood, hence the need for a minimum strength requirement. However, this minimum strength requirement has become entrenched in the standard and is now applied to softwood plywoods as well.

The Japanese and French standards finally have combined the German and the North American concepts of bond evaluation and require both a minimum strength and a minimum percentage of wood failure.

The existence of these conflicting bond requirements can form a real technical trade barrier. This barrier could be removed if all countries could agree on a common standard for testing and evaluating their plywood. Efforts in this direction are made in other areas of the evaluation of wood products, as will be shown later, but not yet in the area of bond evaluation.

DIFFERENCES IN STANDARD DIMENSIONS OF WOOD PRODUCTS

Description of the physical dimensions of wood products is an important part of their specification. To simplify the manufacturing and design processes many countries have standardized the dimensions of lumber and plywood. For example, the standard thicknesses of unsanded construction softwood plywood are given in Table 3 for four countries. Canada decided to keep the number of standard thicknesses to a minimum to avoid inventory problems and reduce the risk of wrong uses of plywoods having only small differences in their thickness. In the United States it was decided to designate a wider range of thicknesses as standard plywood thicknesses.

Table 3. Standard plywood thicknesses, unsanded grades, in Canada, the United States, Germany, and Japan.

Standard	Unit	Standard Thicknesses					
CSA 0121 and 0151	mm	7.5	9.5	12.5	15.5	18.5	20.5
U.S. PS 1-83	inch [mm]	5/16 11/32 [7.9] [8.7]	3/8 13/32 [9.5] [10.3]	1/2 17/32 [12.7] [13.5]	5/8 21/32 [15.9] [16.7]	3/4 25/32 [19.1] [19.8]	7/8 [22.2]
DIN 4078	mm	8	10	12	15	18 20	22
JAS (new)	mm	7.5	9	12	15	18	21

In addition, what is immediately apparent in Table 3 is that the metric equivalents of the fractional inch dimensions seldom are completely identical to the standard thicknesses used in Canada. Also, standard thicknesses in Germany and Japan, with the exception of the 7.5 mm thickness, seem to be whole numbers rather than the decimal fractions in use in Canada. This comparison does not say anything about the tolerances associated with these thicknesses. Thus a Canadian panel which, according to Canadian standards, could be up to 0.5 mm below or 1.0 mm above the nominal thickness, in fact is still acceptable under DIN or JAS standards.

Building practices in any given country have developed around the materials produced in that country. A manufacturer bringing plywood manufactured to U.S. standards into, say, Germany might find it difficult to sell a 13/32 or 17/32 inch panel, because there is simply no market for these panels. This is not to say that it cannot be done. Surely some ingenious marketing expert could find a use for a 13/32 inch panel in, say, Japan; but before he can sell that panel, he has to overcome the invisible technical barrier. The barrier is the JAS standard with its existing range of standard thicknesses.

Consider as another example the standard panel sizes of plywood manufactured in four countries (Table 4). In Canada two panel sizes are recognized in CSA Standards 0121 and 0151. In fact only one size is usually manufactured, 1,220 by 2,440 mm, which is a survivor from the imperial dimensions once used in Canada, the old 4 by 8 foot panel. Apparently, the smaller rational metric panel size is available on request, but more expensive, since it has to be cut from the larger size panel. Again the U.S. standard recognizes a larger range of panel sizes, including the Canadian standard panel size.

Table 4. Standard plywood panel sizes (width and length) in Canada, the United States, Germany, and Japan.

Standard	Dimension and Unit	Standard Sizes									
CSA 0121 and 0151	Width mm	1,220	1,220								
	Length mm	2,400	2,440	(tolerance = + 0, -4 mm)							
PS 1-83	Width inches (mm)	36	(914)	48	(1,219)	60	(1,524)				
	Length inches (mm)	60	(1,524)	96	(2,438)	120	(3,048)				
DIN 4078	Width mm	1,220	1,250	1,500	1,530	...	3,050				
	Length mm	1,220	1,250	...	2,440	2,500	3,050				
JAS (new)	Width mm	900	910	910	910	910	910	955	1,000	1,220	1,220
	Length mm	1,800	1,818	1,820	2,130	2,440	2,730	1,820	2,000	2,440	2,730

The German and the Japanese standards reflect the marketing efforts of North American exporters, since both countries now include the 1,220 by 2,440 mm North American panel size in their standard. But it is still difficult to convince German contractors, for example, to use this size for concrete formwork, since support systems for concrete forms are often designed around the readily available 2,500 mm long panel. Again, standard dimensions can act as a form of trade barrier.

As a result of the use of the old 8 foot plywood panel, ceiling heights in most residences and many commercial buildings in North America today are 8 feet or 2,440 mm. In fact, a whole industry has developed around this 8 foot modular dimension, producing both studs and gypsum wall panels of this length. In Germany a law regulating conditions at the workplace requires the minimum clear height of the workplace to be 2.5 m (Anon. 1986, p. 19). No market there for an 8 foot panel.

In the United Kingdom, truss rafters are usually manufactured using 35.5 mm thick lumber imported from the Scandinavian countries. This dimension is identified in the truss section of BS 5268, the U.K. timber design code, and usually specified by U.K. truss manufacturers. This common practice does affect the sale of North American lumber, which at 38 mm is considered to be too wasteful. However, a reaction has developed against the 35 mm rafter from an unexpected direction. Tradespeople responsible for nailing plasterboard to this narrow lumber found that it was difficult to nail the edges of the panels to a 35 mm wide joist where the panels are butting against each other. Either the nail would miss the lumber altogether or the wood would split. A plasterboard applicators standard is therefore being developed, which if accepted, would make a minimum lumber thickness of 38 mm mandatory for constructions involving plasterboard. But such a standard would certainly be viewed as a technical trade barrier by Scandinavian lumber exporters.

THE BENDING STRENGTH OF PLYWOOD

Today as a result of the ready availability of electronic computers, it is fashionable to talk about modeling physical or other phenomena. By modeling is meant the attempt to describe reality in terms of some mathematical relationship. Because these relationships often involve many unknown quantities, the resulting mathematical models are rather complex, and it would be practically impossible to solve the resulting systems of equations without the aid of high speed computers.

However, model building in one form or another has been practiced by engineers for a long time. The French mathematician M. H. Navier (1785-1836) suggested that the deformations that would develop in a beam subjected to lateral loads would be such that a cross section that was originally plane would remain so during bending. Another assumption made by engineers is that as a result of external forces applied to a beam, internal stresses develop. These purely hypothetical stresses have to obey Newton's law that every action will require a reaction (i.e., internal and external forces have to be in equilibrium).

The model based on these two assumptions, the equilibrium of forces and the maintaining of plane sections during bending, leads to the hypothesis that for an elastic material bending stresses will vary linearly across a section from a maximum compressive stress on the outside face of the concave side of the beam to a maximum tensile stress on the opposite beam face. In mathematical language the model can be stated as:

$$M = f \, I/c$$

where   M   =   bending moment
         f   =   stress at upper or lower face of a beam
         I   =   second moment of area of cross section about its neutral axis
         c   =   distance of outer faces from neutral axis

This model has been used widely for the evaluation of tests on beams of various materials, including wood, as well as for their analysis and design. The model has some shortcomings, the most obvious being that most materials do not remain elastic right up to failure (i.e., strains and stresses are not always proportionate). Nevertheless, this model of beam bending has been extremely useful in the interpretation of many tests on wooden bending members.

The model also is used to interpret the results of bending tests with plywood. And here we run into two quite different uses of this model that result in some apparent contradictions which are carried right through into various national standards.

Consider the cross section of a strip of plywood (Figure 4). The layers of veneer in the plywood are placed in such a way that the wood fibers in adjacent layers are at right angles to each other. Since the strength and stiffness of wood in the longitudinal direction are many times that of wood in its tangential direction, we have a succession of strong and weak layers. In a sense the orientation of veneer layers at right angles provides a reinforcing effect for the weak direction of each layer, and we end up with a sheet of plywood that has a more uniform strength and stiffness in the major directions than we would have if all the veneer layers were oriented in the same direction.

To find the bending strength of this plywood, we cut a narrow strip of material from a plywood panel. We support this strip of plywood at its ends, apply a lateral load at the center, and increase this load until the plywood breaks. From the breaking load and the span of the supports for the specimen, we can calculate the bending moment at failure. Navier's mathematical model now states that the maximum stress at the surface of the plywood, the so-called modulus of rupture, is equal to the ratio of the bending moment at failure and the section modulus $S = I/c$ of the beam cross section.

The bending moment, of course, has been calculated from the experimentally observed maximum load and the test span. But how is the section modulus to be calculated? Two quite different approaches have been developed for this. In North America it is argued that only those plies that

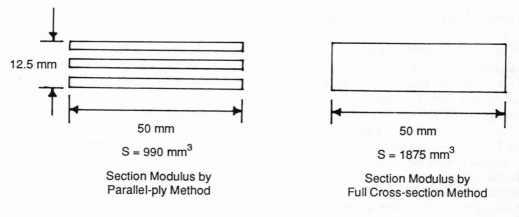

Figure 4. Plywood section modulus calculations.

are oriented so that their grain runs parallel to the span are actually carrying the load, because the cross plies are so flexible and weak that they will not contribute to the load-carrying capacity of the section. On the basis of this assumption a second moment of area, I, and a section modulus, S, for the plywood cross section can be calculated. This is the so-called parallel-ply method of analysis and will lead to certain values for the modulus of elasticity and the modulus of rupture.

In most other industrialized countries engineers have put forward the argument that they are not really interested in the internal construction of plywood and that they are quite content to assume that the plywood is homogeneous throughout its depth. This is the so-called full-section method of analysis. This assumption of homogeneity will result in a different second moment of area for the same plywood cross section and consequently a different modulus of elasticity and modulus of rupture.

The ASTM standard used in the United States recognizes that there is more than one way the second moment of area can be calculated (ASTM D 3043, 1976), but in actual practice only the parallel-ply method of analysis is being used (APA 1983). Both the German (DIN 68 705, 1981) and the Japanese (JAS, new) standards assume the full-section method of analysis. The resulting numerical values for the strength and modulus of elasticity of plywood by the full-section method of analysis are roughly half of those calculated using the parallel-ply method.

The parallel-ply method of analysis has the advantage that it allows quite readily the calculation of the moment carrying capacity and stiffness of plywood lay-ups that have not actually been tested. It therefore is a method favored in a plywood producing country such as the United States. On the other hand, the full-section method fits more readily into the system of design assumptions made by structural engineers who have to design not only with plywood but with many other structural materials such as steel, aluminum, or plastics. The full-section method therefore is used by many European countries as well as Japan where more emphasis is placed on the structural design of all structures, including small buildings and residential structures.

One solution to this apparent dilemma is not to specify allowable stresses at all, but rather to tabulate the moment resistance for various thicknesses of plywood. This is done in both the Danish (SBI 135, 1983) and the latest Canadian design standard for wood structures (CSA 086.1-M84).

The question of bending strength is further complicated by the size of test specimens used to determine the fundamental bending properties of plywood. In Germany, for example, specimens of small size, only 50 mm wide, are being used. Such small specimens measure the basic strength of plywood quite well, but do not account for the presence of knotholes and other imperfections in construction type plywood. Allowances for these have to be made by modifying the test data in some rational manner.

In North America, on the other hand, large size panels are used to determine the bending capacity of plywood. The sizes of these test panels are large enough to be representative of plywood as it is being used in construction. No further modifications of the test data to account for possible imperfections are required.

The Japanese plywood standard recently was modified to allow for the testing of large size panels, and comparisons of strength values for plywood determined in Canada, the United States, and Japan are therefore now possible. The same cannot be done for data determined in Germany.

CODES AND STANDARDS FOR THE DESIGN OF ENGINEERED STRUCTURES IN WOOD

In the previous sections, standards for the description of measuring units and standards for some forest products and their testing have been considered. This section will consider codes and standards for structural design with wood products.

Structural design is the process of assigning dimensions to the various parts of a planned structure such that, when the structure has been built, it will be capable of safely performing its intended functions. Like any other human activity, such a design process involves a certain amount of risk.

To ensure that this risk is kept at an acceptable level, structural engineers have agreed to use certain standard procedures in the design of engineered structures. These standards may include assumptions that have to be made about the type and magnitude of forces that have to be assumed in the design. They also include the safe resistance to the stresses generated by these forces which can be expected from various materials. Where these standards are incorporated into a framework of legal, public documents, these standards are referred to as codes.

Typical structural engineering design standards are CSA O86-M84, a standard on the Engineering Design in Wood published by the Canadian Standards Association, and the National Design Specification for Wood Construction prepared by the National Forest Products Association (NFPA) in the United States. Since the whole of CSA O86 is incorporated into the National Building Code of Canada (NBCC) and the NBCC has been adopted as a legal document by all the provinces in Canada, CSA O86 also is the design code that in Canada legally determines the design of all building construction in wood, where the building area exceeds 600 square meters or three stories in height. Similar design standards and codes exist in all industrialized countries.

Complementing these design standards are standards for the design of small buildings. Such standards often specify simple rules for the safe use of the most common standardized building materials available in a country. These rules allow the selection of materials for the construction of small buildings by builders or architects without reference to a structural design engineer. For example, Part 9 of the National Building Code of Canada contains such rules which apply to all buildings not exceeding 600 square meters in area or three stories in height.

No universal agreement exists about the level of safety for which a structure should be designed. The issue of safety furthermore is clouded by the lack of agreement on how to assess safety. This area is one in which much research is currently being carried out.

As an example of the lack of agreement between national standards, the Canadian standard CSA O86 and the corresponding British design standard BS 5268 should be mentioned. In their so-called working stress format both standards generally start out with the same data base for the strength of lumber. But the stresses allowed by the British standard are lower than those allowed by the Canadian standard because the latter assumes a factor of safety of 2.1, whereas the British standard assumes a factor of safety of 2.3. Again these numbers cannot be compared directly because a complete assessment would have to consider the basis for the load assumptions made in both countries.

A consideration of safety is not the only factor taken into account during design. Serviceability and the general perception that people have about the quality of the performance of a structure are important factors that have to be considered by the designer. To assess the performance of floors, standards have been written such as the section on performance testing in U.S. PS 1-83.

This performance standard limits the deflection of plywood used for floor construction under certain test conditions to not more than 6.4 mm. An almost identical test used in Scandinavian countries sets the level of acceptable deflection at 2.5 mm. This example illustrates the difficulty manufacturers have to face when they are marketing abroad a product acceptable for a certain end use in their own country.

HARMONIZATION OF NATIONAL STANDARDS

The preceding examples have illustrated some of the problems created by the existence of different national product, testing, and design standards. The important role in international trade played by technical standards has been recognized by the General Agreement on Tariffs and Trade (GATT): "Gatt is a multilateral treaty, subscribed to by 90 governments which together account for more than four-fifths of world trade. Its basic aim is to liberalize world trade and place it on a secure basis, thereby contributing to economic growth and development and the welfare of the world's people" (GATT 1986).

During the Tokyo Round of multilateral trade negotiations, concluded in November 1979, ninety countries including Canada and the United States agreed to ensure that national technical standards, testing, and certification schemes related to them would not be allowed to create unnecessary obstacles to trade. This particular section of the GATT is generally known as the Standards Code. In the field of building construction there are two international organizations that are contributing substantially to the implementation of the principles formulated by the GATT.

To resolve some of the problems that arise from differences in national design standards, the International Council for Building Research, Studies, and Documentation (CIB) some years ago started a process of harmonizing design standards. For timber design, CIB's committee W18 has prepared a model code that has formed the basis of Eurocode 5, Common Unified Rules for Timber Structures. It is the intention of the European Economic Community to allow design of timber structures based on Eurocode 5 alongside those designs based on the various national codes of member countries of the European Community. It is obvious that such a step will remove one invisible barrier to the free flow of engineering services and prefabricated timber structures within the European Community.

On a wider level the International Standards Organization (ISO), which is active in many areas of standardization, also has initiated the preparation of a timber engineering design standard. ISO Technical Committee 165 is currently preparing a design document that will serve as a model for all countries participating in the work of this international body. Again the design code prepared by CIB W18 serves as a guide for the preparation of this model code.

CONCLUSIONS

The concept of standards has guided building activities for several thousand years. Standards are based on the common experiences of groups of people. It is therefore not surprising that measuring, material, and building standards will differ from country to country. In an age of widely increased international trade, such differences in standards can constitute an invisible barrier to trade in wood products, as has been recognized by the GATT. If we believe that international trade can be beneficial to the economies of both importing and exporting countries and if we want to live up to the spirit of the Standards Code of the GATT, then every effort should be made to harmonize product and building standards on an international level. Early implementation of the SI system of metric measurements by those few countries that have not yet adopted it will go a long way toward realizing the goal of common product and building standards.

REFERENCES

Anonymous. 1986. Bausortiment und Heimwerker-
bedarf, October 31, p. 19.

Engel, H. 1964. The Japanese house. Charles E. Tut-
tle Co., Tokyo.

GATT Information Service. 1986. Information
leaflet, Agreement on Tariffs and Trade: What it
is, what it does.

Klein, H. A. 1975. The world of measurements.
Simon and Schuster, New York.

Skinner, F. G. 1967. Weights and measures: Their
ancient origins and their development in Great
Britain up to A.D. 1855. Her Majesty's
Stationery Office, London.

STANDARDS CITED

APA. 1983. Plywood design specification. American
Plywood Association, Tacoma, Washington.

ASTM D 3043. 1976. Standard methods of testing
plywood in flexure. American Society for Testing
and Materials, Philadelphia.

ASTM E 380. 1984. Standard for metric practice.
American Society for Testing and Materials,
Philadelphia.

BS 5268: Part 2: 1984. Structural use of timber.
British Standards Institution, Hemel Hempstead,
Herts., United Kingdom.

CIB Publication 66. 1983. Structural timber design
code. Working Group CIB W18. International
Council for Building Research, Studies, and
Documentation.

CSA 086.1-M84. Engineering design in wood (limit
states design). Canadian Standards Association,
Rexdale, Ontario.

CSA 0121-M78. Douglas-fir plywood. Canadian
Standards Association, Rexdale, Ontario.

CSA 0151-M78. Canadian softwood plywood.
Canadian Standards Association, Rexdale, Ontario.

DIN 4078. 1979. Sperrholz-Vorzugsmasze.
Deutsches Institut für Normung.

DIN 53 255. 1964. Bestimmung der Bindefestigkeit
von Sperrholzleimungen. Deutsches Institut für
Normung.

DIN 68 705. 1981. Sperrholz. Teil 3. Deutsches Institut für Normung.

Eurocode 5. 1986. Common unified rules for timber
structures report. Danish Building Research Institute, Copenhagen.

JAS 894. 1977. Japanese agricultural standard for
structural plywood. Japan Plywood Inspection
Corporation, Tokyo.

JAS (new). Japanese agricultural standard for structural plywood. Japan Plywood Inspection Corporation, Tokyo.

NBCC. 1980. National building code of Canada. National Research Council of Canada.

NDS. 1986. National design specification for wood
construction. (U.S.) National Forest Products Association, Washington D.C.

NFB 51-338. 1978. Contreplaqué à plis. Plans de col-
lage. Méthodes d'essais. L'Association Française
de Normalisation (AFNOR).

Placos 894. 1977. Panneaux contreplaqués. CTB.

PS 1-83. 1983. U.S. product standards for construction and industrial plywood. U.S. National
Bureau of Standards, Washington, D.C.

SBI 135. 1983. Trackonstruktioner—Beregning.
Statens Byggeforskningsinstitut.

Chapter 15

Impact of Building Codes on Trade in Forest Products in Southeast Asian and Pacific Rim Countries

R. H. LEICESTER

In July 1984 the Australian Standard AS 2878 Timber—Classification into Strength Groups (Standards Association of Australia 1986a) was complete and ready for postal ballot prior to publication. At this final stage, New Zealand's Minister for Overseas Trade and Marketing successfully intervened to halt the course of the Standard. This move was initiated by the New Zealand Timber Industry Federation, which had noted that in the new Standard the proposed strength groupings for New Zealand radiata pine and Douglas-fir would have adversely affected the export trade to Australia. It was claimed that at best the Standard would cost the exporters an additional U.S.$3 million per year; at worst it could drastically reduce the trade.

During the past decade there have been numerous examples of the impact (or potential impact) of building Standards on the trade of forest products imported into Australia. Usually these Standards have been used by the Australian forest products industry to maintain product quality in the marketplace. But there is little doubt that these same Standards could be used to provide the means for the indigenous forest products industry to resist incursions into their share of the Australian market. Roughly one-third of Australia's sawnwood consumption (as indicated in Table 1) and approximately half of its plywood consumption are made up of imports; such a situation must obviously lead to market tensions. The following examples give some further idea of the negative impact of building Standards on trade into Australia.

Table 1. Annual sawnwood consumption in Australia.

Source	Consumption* (1,000 m³)
Indigenous timber	
Eucalypts	1,729
Plantation conifers	1,099
Other	170
Imported timber	
Conifers	1,079
Hardwoods	297

*From Bureau of Agricultural Economics (1986). Roughly 50% of sawn timber consumption is controlled by building Standards.

In 1981 imports of North American spruce-pine-fir for construction purposes were successfully resisted with the argument that at the time the visual grading of such material was not covered by any Australian Standard. In 1982 building Standards were again used by the indigenous industry, this time as part of a battle against the alleged dumping of New Zealand radiata pine into Australia. In 1985 the strength classifications proposed in AS 2858 Timber-Softwood-Visually Stress-Graded for Structural Purposes (SAA 1986b) for North American Douglas-fir were seen as such a major threat to the import trade from North America that a top ranking delegation of North American industry leaders and researchers was flown out to Australia to argue the case for a more favorable strength classification.

In addition to their impact on sawnwood imports, Australian building Standards have had an effect on imported composite products. New Zealand has had continual trouble in the export of scaffold planks to Australia because it is difficult to establish the equivalence between the relevant New Zealand Standard NZS 3620 (Standards Association of New Zealand 1985) and the corresponding Australian Standards AS 1577 and AS 1578 (SAA 1974a, 1974b).

In 1986 the Plywood Association of Australia successfully initiated a court action to prevent plywood from an importing consortium from being sold in Australia as concrete formwork. The plywood was fabricated from many Southeast Asian species, and the argument here was that there was inadequate control over the choice of species used for the veneer. Although the timber was marketed as plywood with an F17 stress grade, a random check on a small sample of the product indicated that when evaluated according to the Australian Standard AS 2269 (SAA 1979), it was probable that the imported plywood sheets would have stress grades ranging from F7 to F22.

Building Standards, if suitably framed, can be of considerable assistance to the trade of forest products, even to the extent that they will open up new types of markets for raw materials.

An example of this is to be found in the format of Australian Standards that are written in terms of structural groups. These structural groups lead directly to building designs as illustrated in Figure 1. Thus if a population of imported sawnwood is classified into a structural group, it can be immediately used for building construction in Australia. This holds true even if the timber had not previously been used for such a purpose.

It is of interest to note that currently there are about 1,500 species of world timbers that have been strength grouped according to the Australian system mentioned above (Berni et al. 1979, Bolza and Keating 1972, Keating and Bolza 1982, SAA 1986a). Potentially all of these species may be exported and used immediately in any country that has building Standards that accept this strength grouping system.

The effectiveness of such a system is indicated by the fact that UNIDO has applied it in the design of a modular bridge system for developing countries (UNIDO n.d. and 1985). The relevant UNIDO design manual lists almost 100 species from the Pacific Rim and Southeast Asian region, most of which have never been used previously for bridge building. In another project, UNIDO has also used the system for the classification of sawnwood from coconut timber, a monocotyledon (Sulc 1983).

SOUTHEAST ASIAN AND PACIFIC RIM COUNTRIES

Multiple Species

For the indigenous forests of Southeast Asia and the Pacific Rim countries, their one natural characteristic that forms the greatest difficulty in trade and utilization is the occurrence of multiple species. In the rain-forest areas of northeastern Australia, each mill has a "compulsory list" of 120 species that it must accept for conversion; Pong Sono (1974) has listed approximately 200 species of merchantable timber in the indigenous forests of Thailand; Espiloy (1977, 1978) notes that in the Philippines there are over 3,000 species of timber of which several hundred are potentially merchantable; Wong and Wong (1980) mention the marketing of 650 species in Malaysia.

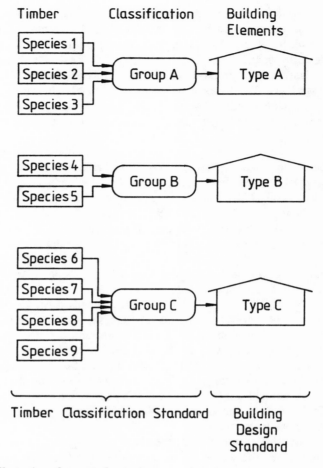

Figure 1. Schematic illustration of structural grouping.

Even when the timber is used within the country of origin, it is not feasible to produce building designs for each grade of each species of timber. In practical terms, total timber use can be obtained only within the framework of a structural group classification system such as that indicated schematically in Figure 1. Apart from Australia, such systems are used in the Southeast Asian and Pacific Rim countries by Fiji (Anon. 1968, 1970), Indonesia (Suparman Karnasudirdga et al. 1978, Abdurahim Martawijaya and Iding Kartasujana 1981), Laos (Timber Research and Development Association 1976), Malaysia (Engku Abdul Rahman bin Chik 1972, Burgess 1956), Papua New Guinea (PNG Department of Forests 1972, Eddowes 1977), the Philippines (Espiloy 1977, 1978), Singapore (Singapore Timber Standardisation Committee 1966), and the Solomon Islands (Forestry Division 1976, 1979).

In theory, these classification systems could be used for importing sawn timber into any particular country. However, because most of these systems differ from one country to another, they are not as useful for trade as they could be.

Apart from the problem of selection of a workable format, a perennial difficulty with the utilization and trade of multiple species forests is the problem of species identification. Identification of the standing tree is difficult enough, but identification of sawn timber is frequently impossible without the use of extremely sophisticated laboratory techniques. Within Australia, a technique termed "proof grading" has been developed which copes with this problem (Leicester 1984,

1985a). In this technique (now used in some twenty mills) a proof load is applied to every stick of timber to ensure that it has at least some of the structural properties required of the claimed stress grade. Although the technique is quite effective, it is not yet recognized internationally, and even within Australia there is at present only a draft Standard (SAA 1986c).

Difficulties in Trade

When trade is influenced by building Standards, the Southeast Asian and Pacific Rim countries are disadvantaged by the fact that most building Standards, including those of the International Standards Organization (ISO), are modeled after existing European and North American Standards. Those Standards are directed toward the efficient utilization of a few softwood species, whereas most Southeast Asian and Pacific Rim countries are concerned with a wide range of multiple species hardwood timbers.

Some of the disadvantages of this situation are obvious. For example, visual grading techniques that are so effective for the classification of softwoods do not work so well with hardwoods, where the critical feature is not usually a knot but the slope of the wood grain—a feature that is difficult to detect in dark colored timbers, particularly if they are rough sawn. The influence of characteristic hardwood defects, such as phloem and kino veins, have not been adequately researched; and for monocotyledon species, such as coconut wood and bamboo, there are no suitable North American or European Standards.

There are also numerous disadvantages of Standards developed in Europe and North America that are not immediately obvious. For example, requirements that classifications be based on tests undertaken at a temperature of 20°C means that expensive air-conditioning costs must be incurred by tropical countries. Similarly a requirement once suggested, that a standard size of 200 mm wide timber be used for sawnwood testing, would be difficult to fulfill in countries that do not usually cut such large sizes in their commercial operations.

Perhaps the most subtle difficulty that could be encountered by tropical countries is that associated with the amount of testing required for classification. In Europe and North America, where there are only a few commercial species with similar properties, there is a good marketing reason for pursuing a high efficiency in the classification of sawnwood and other forest products. This sort of high efficiency testing for the sawnwood of just a *single* species requires the testing of thousands of pieces of structural size timber, perhaps even tens of thousands, and would be expected to absorb the resources of a good sized laboratory for at least one year. Obviously such an assessment procedure is not possible for countries with limited laboratory facilities or with hundreds of species to utilize.

What is required is a classification system whereby a trade-off can be made between efficiency and testing costs. Frequently a loss in efficiency is of little consequence for hardwood timbers, because their basic structural properties may be far superior to that of construction softwoods, as illustrated in Figure 2.

DISCUSSION ON BUILDING STANDARDS

General

In this section, Australian building Standards for sawnwood will be used as a basis for discussing the drafting of Standards to assist international trade. The forest products industry in Australia, with its softwood plantations, temperate zone eucalypt hardwood forests, and tropical zone rain forests, displays in a compact form the characteristics of the timber industries of the world. Because of the great diversity in forest types, the concept of structural grouping was introduced into the format of Australian Standards more than fifty years ago.

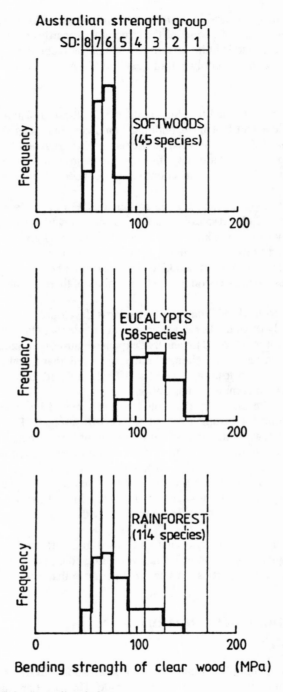

Figure 2. Range of strength for Australian timbers.

Australian Standards for Sawnwood

Australian engineering Standards are stated in terms of design properties rather than properties of the timber. For example, in the Timber Engineering Code AS 1720 (SAA 1975), design properties for sawnwood that are related to clear-wood properties are stated in terms of a strength group classification; design properties related to natural defects are stated in terms of a stress grade; and design properties of connector systems are stated in terms of a joint group. Examples of these classifications are given in Tables 2, 3, and 4.

Table 2. Design property for strength groups.

Strength Group	Compression Perpendicular to the Grain (MPa)
SD1	10.4
SD2	9.0
SD3	7.8
SD4	6.6
SD5	5.2
SD6	4.1
SD7	3.3
SD8	2.6

Source: SAA (1975).

Table 3. Design properties for stress grades.

Stress Grade	Basic Bending Strength (MPa)	Modulus of Elasticity (MPa)
F34	34.5	21,500
F27	27.5	18,500
F22	22.0	16,000
F17	17.0	14,000
F14	14.0	12,000
F11	11.0	10,500
F8	8.6	9,100
F7	6.9	7,900
F5	5.5	6,900
F4	4.3	6,100
F3	3.4	5,200
F2	2.8	4,500

Source: SAA (1975).

For trade purposes, the important matter is the method whereby timber is classified into these design classes. For the classification into strength groups (Table 5), data from standard mechanical tests on small clear pieces of wood are required (Mack 1979), and if this is not available, a conservative classification may be made on the basis of density information only (SAA 1986a). For classification into joint groups (Table 6), only density information is required (Mack 1978, Leicester and Keating 1982).

Table 4. Design property for joint groups.

Joint Group	Basic Lateral Load on 3.15 mm Diameter Nail (N)
J1	455
J2	390
J3	330
J4	230

Note: The joint groups will be extended in the revision of AS 1720.
Source: SAA (1975).

Table 5. Classification criteria for strength groups.

Strength Group	Method I: Bending Strength of Small Clear Wood Specimens at 12% Moisture Content (MPa)	Method II: Density at 12% Moisture Content (kg/m^3)
SD1	150	1,200
SD2	130	1,080
SD3	110	960
SD4	94	840
SD5	78	730
SD6	65	620
SD7	55	520
SD8	45	420

Source: SAA (1986a).

Table 6. Classification criterion for joint groups.

Joint Group	Basic Density (kg/m^3)
J1	750
J2	600
J3	475
J4	380

The stress grade classification depends on the method of grading used. In the case of visual grading, the timber is first given a percentage grade, termed a structural grade, that depends on the size of the observed natural defects such as knots, slope of grain, pith, and checks (SAA 1977, 1986d). The stress grade is then derived from the structural grade and strength group as shown in Table 7.

Table 7. Classification criterion for stress grades.

Visual Grade*		Stress Grade							
Nomenclature	% strength of clear material	SD1	SD2	SD3	SD4	SD5	SD6	SD7	SD8
Structural grade no. 1	75		F34	F27	F22	F17	F14	F11	F8
Structural grade no. 2	60	F34	F27	F22	F17	F14	F11	F8	F7
Structural grade no. 3	48	F27	F22	F17	F14	F11	F8	F7	F5
Structural grade no. 4	38	F22	F17	F14	F11	F8	F7	F5	F4

*Measured according to AS 2082 and AS 2858 (SAA 1977, 1986d).

These classification tests are prescriptive and as such do not provide a direct measurement of the required structural properties. When more accurate classifications are required, such as in cases of dispute or when greater material efficiency is sought, classifications based on performance criteria should be used. Although these tests cannot fully simulate in-service conditions, they endeavor to give the best approximation possible from a laboratory test.

In Australia, performance tests for some of the design properties given in Tables 2, 3, and 4 can be found in publications by the SAA (1974c, 1986b).

As an example of a performance specification, consider the method of measuring the in-service bending strength. To do this, a sample of representative beams are selected and loaded in third point bending as shown in Figure 3. The basic stress in bending F_b to be used for design is given by

$$F_b = F_{b,0.05} (\eta/\gamma\xi) \qquad (1)$$

in which

$$\eta = 1 - (2.7\, V_b/\sqrt{N}) \qquad (2)$$
$$\gamma = 1.75 \qquad (3)$$
$$\xi = 1.3 + 0.7\, V_b \qquad (4)$$

where $F_{b,0.05}$ is the five-percentile value of the measured strength, V_b is the coefficient of variation of strength, $N \geq 30$ is the sample size, η is a factor to compensate for the uncertainty involved in the use of small sample sizes, ξ is the ratio between short-term and basic strength, and γ is a factor of safety.

A point to note is that the parameter η may be used to assess the effects of a trade-off between design efficiency and cost of testing.

Difficulties

The Australian Standards are representative of the more developed Standards, and yet they present difficulties for international trade.

If prescriptive Standards are used, it is usually difficult to compare forest products that have been classified in two different countries. Design properties cannot be compared because the implied factors of safety are unknown; they will depend to some extent on the design loads specified in the national loading Standards. Prescriptions are not easily compared; for example, in the measurement of knot sizes in visual grading two different methods may be used (Figure 4).

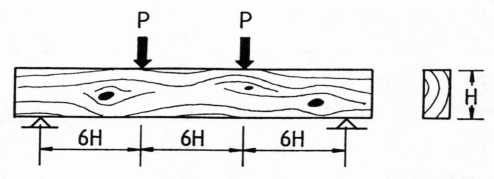

Figure 3. Standard test to measure bending strength. (Members are selected so that natural defects occur at random.)

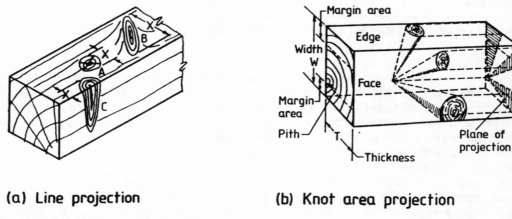

(a) Line projection

(b) Knot area projection

Figure 4. Methods for measuring knot size.

On the other hand, performance Standards are extremely expensive to implement, and the costs and time involved are difficult to cover in the introduction of a new product. Moreover, it turns out that even performance-based standards do not produce unique classifications, and require some form of international agreement for their implementation. For example, the nominal bending strength of timber varies with the way in which it is loaded, the span-to-depth ratio of the beam, and the selection of the specimen with respect to the defect locations (Leicester 1985b). There is no international agreement on these test parameters.

It is worth mentioning that materials classified for use with limit states codes, the next generation of building Standards for structural products, will be specified in terms of *characteristic* values that are essentially performance properties. An example of this is given in the draft Eurocode, a model code for timber engineering (Crubilé et al. 1986). It is hoped that common definitions of characteristic values will be adopted throughout the world.

BUILDING STANDARDS TO ASSIST TRADE

International Standardization

The fundamental problem that links building Standards with international trade is that of classifying a product in one country in such a way that it will be acceptable in another. Thus it is apparent that standardization must have a common, or at least a comparable, basis in all countries if national Standards are not to act in some degree as a barrier to trade. There are several ways by which this may be attempted, the most obvious being through the common adoption of international Standards. The logical Standards to use for this purpose are those currently under development by ISO.

To obtain international acceptance, the ISO Standards will have to be global in their applicability, without any bias toward any of the current trading patterns. If for no other reason, it is prudent to avoid a bias because this will involve difficult technopolitical maneuvers as trade patterns change. There is of course no reason why some particular localized trading group (such as Australia and New Zealand) may not develop a special set of temporary Standards for themselves, should the countries involved so desire.

International Standards should cover (without bias) the full range of physical parameters associated with the production and use of forest products; these parameters include timber species, climate, and wood-destroying life forms. But care should be taken to ensure that the full range of human parameters are also covered; these include technology levels, financing available, and laboratory and quality control facilities. An example of this would be to build in the facility to enable a trade-off between testing requirements and efficiency in building product classification.

Prescriptive and Performance Criteria

Experience indicates that for the purpose of assisting trade, all Standards, whether local or international, should include at the very least a statement of the intended target performance. This is required to settle cases of dispute related to compliance with a prescriptive Standard. Ideally all Standards should describe some sort of test that will give a good simulation of in-service performance.

An extreme example of prescriptive Standards that restrict trade is to be found in the British Standard 1990-1984 Wood Poles for Overhead Power and Telecommunication Lines (British Standards Institution 1984) and the American Standard ANSI 05.1-1979, American National Standard Specifications and Dimensions for Wood Poles (American National Standards Institute 1979). Both Standards specify the species that are acceptable for compliance; they do not include radiata pine, for example, as an acceptable species even though it probably has the required performance characteristics; as a result, New Zealand has not been able to tender for pole supply contracts in the United Kingdom and the United States.

Unfortunately, the use *solely* of performance criteria for the classification of forest products is an ideal that is not practicable for several reasons. Some performance targets, such as durability or long-term strength, are not always possible to evaluate through short-term tests. In addition, some tests, such as the tests mentioned earlier for bending strength, are not feasible if they have to be applied to every size of timber of every grade of every species where numerous species are involved. The same type of difficulty applies to performance tests for connector systems.

From these considerations it appears that for purposes of international trade two types of international Standards are required: (1) performance Standards that measure as closely as possible in-service performance characteristics; (2) deemed-to-comply prescription Standards that may not be very efficient but are nonetheless simple and inexpensive to apply.

Grouping

As indicated earlier, international trade is assisted by the use of group classifications. Grouping is of particular assistance to new products, because through use of this concept new products need be involved only with product classification and not building design as shown in Figure 1.

An interesting question yet to be discussed is the choice of property to be grouped. Traditionally, as in the example given for Australian Standards, the grouping is applied to material properties such as strength per unit area. While this is adequate for trade purposes, a case could be made for the grouping of total structural elements such as floor systems, trusses, and shear walls. In this way not only is the building material traded but also the building technology that was applied in the development of the elements. To some extent this has already occurred with the trade of plywood, glulam, and scaffold planks.

CONCLUSION

In the past, national building Standards have played roles both in assisting the international trade of forest products and in acting as obstacles to that trade. For many Southeast Asian and Pacific Rim countries special difficulties arise because of the multiple species characteristic of the indigenous forests.

It is argued that to assist in the international trade of forest products it is essential to have international standardization of national building Standards. These Standards need to be unbiased toward any particular trading pattern, to be sufficiently flexible to cope with both high and low technology infrastructures, and to cope with both single species softwood forests and multiple species hardwood forests.

For global effectiveness and in particular to satisfy the special needs of Southeast Asian and Pacific Rim countries, both performance and prescriptive Standards are required. These Standards should be written in terms of grouped classifications rather than species-specific ones. (This would not prevent the use of species-specific Standards within a localized trade association.) In addition to Standards for materials, consideration should be given to the development of Standards for total building elements.

ACKNOWLEDGMENTS

The author is indebted to W. G. Keating (CSIRO), G. B. Walford (Forest Research Institute, New Zealand), and K. G. Lyngcoln (Plywood Association of Australia) for helpful discussions in the preparation of this paper.

REFERENCES

Abdurahim Martawijaya and Iding Kartasujana. 1981. The potential use of Indonesian timbers. Industrial Agricultural Research Development Journal 3(4):108-116.

American National Standards Institute. 1979. ANSI 05.1-1979. American National Standard specifications and dimensions for wood poles. ANSI, New York, N.Y.

Anonymous. 1968. Some timbers of Fiji. Department of Forestry, Suva.

——. 1970. Fiji timbers and their uses. Leaflet Series. Department of Forestry, Suva.

Berni, C., E. Bolza, and F. J. Christensen. 1979. South American timbers: The characteristics, properties and uses of 190 species. CSIRO Australia, Division of Building Research Melbourne.

Bolza, E., and W. Keating. 1972. African timbers: Properties, uses and characteristics of 700 species. CSIRO Australia, Division of Building Research Melbourne.

British Standards Institution. 1984. British Standard BS 1990-1984. Wood poles for overhead power and telecommunication lines. Part 1. Specification for softwood poles. BSI, Hemel Hempstead, Herts., United Kingdom.

Bureau of Agricultural Economics. 1986. Forest products. Timber Supply Review 36(3):1-17.

Burgess, H. J. 1956. Strength grouping of Malayan timbers. Malayan Forester 19(1):33-36.

Crubilé, P., J. Ehlbeck, H. Brüninghoff, H. J. Larsen, and J. Sunley. 1986. Eurocode 5: Common unified rules for timber structures report. Danish Building Research Institute, Copenhagen.

Eddowes, P. J. 1977. Commercial timbers of Papua New Guinea: Their properties and uses. Forest Products Research Centre. Hebams Press, Port Moresby, PNG.

Engku Abdul Rahman bin Chik. 1972. Basic and grade stresses for strength of groups of Malaysian trees. Malayan Forester 35(2):131-134.

Espiloy, E. B. 1977. Strength grouping of Philippine timber species for structural purposes. NSDB Technology Journal, October-December, pp. 76-86.

——. 1978. Strength grouping of Philippine timbers for utilisation of lesser-known species. Tech. Note 187. Forest Products Research and Industrial Development Commission, National Science Development Board College, Laguna 3720, Philippines.

Forestry Division, Ministry of National Resources. 1976. Solomon Island timbers. Timber Booklet 1: Major species. Government Printing Office, Honiara.

——. 1979. Solomon Island timbers. Timber Booklet 2: Minor species. Government Printing Office, Honiara.

Keating, W. G., and E. Bolza. 1982. Characteristics, properties and uses of timbers. Vol. 1: South-East Asia, Northern Australia and the Pacific. Inkata Press, Melbourne.

Leicester, R. H. 1984. Proof grading. Proceedings of Pacific Timber Engineering Conference, IPENZ, Auckland, New Zealand, 3(May):768-778.

——. 1985a. Proof grading technique. Paper 3-11. Timber Engineering Symposium on Forest Products Research International—Achievements and the Future, Pretoria, South Africa.

——. 1985b. Configuration factors. Paper 3-8. Timber Engineering Symposium on Forest Products Research International—Achievements and the Future, Pretoria, South Africa.

Leicester, R. H., and W. G. Keating. 1982. Use of strength classifications for timber engineering standards. CSIRO Australia, Division of Building Research Tech. Pap. (Second Series) 43.

Mack, J. J. 1978. The grouping of species for the design of timber joints with particular application to nailed joints. CSIRO Australia, Division of Building Research Tech. Pap. (Second Series) 26.

——. 1979. Australian methods for mechanically testing small clear specimens of timber. CSIRO Australia, Division of Building Research Tech. Pap. (Second Series) 31.

Papua New Guinea Department of Forests. 1972. New horizons, forestry in Papua New Guinea. Jacaranda Press, Brisbane.

Pong Sono. 1974. Merchantable timbers of Thailand. In Symposium on Research and Marketing of South-East Asian Timbers and Timber Products, organized by the German Foundation for International Development (Berlin) and the Government of the Republic of the Philippines, Manila and Los Banos, Philippines, pp. 120-165.

Singapore Timber Standardisation Committee. 1966. Code of practice for the use of timber in building construction. Singapore Industrial Research Unit, Singapore.

Standards Association of Australia. 1974a.

Australian Standard AS 1577-1974. Solid timber scaffold planks. SAA, North Sydney.

——. 1974b. Australian Standard AS 1578-1974. Laminated timber scaffold planks. SAA, North Sydney.

——. 1974c. Australian Standard AS 1649-1974. Determination of basic working loads for metal fasteners for timber. SAA, North Sydney.

——. 1975. Australian Standard AS 1720-1975. SAA timber engineering code (metric units). SAA, North Sydney.

——. 1977. Australian Standard AS 2082-1977. Visually stress-graded hardwood for structural purposes. SAA, North Sydney.

——. 1979. Australian Standard AS 2269-1979. Structural plywood. SAA, North Sydney.

——. 1986a. Australian Standard AS 2878-1986. Timber—Classification into strength groups. SAA, North Sydney.

——. 1986b. Draft Australian Standard for the evaluation of strength and stiffness of graded timber DR 83205. SAA, North Sydney.

——. 1986c. Draft Australian Standard for proof grading timber. SAA, North Sydney.

——. 1986d. Australian Standard AS 2858-1986. Timber-softwood-visually stress-graded for structural purposes. SAA, North Sydney.

Standards Association of New Zealand. 1985. New Zealand Standard NZS 3620-1985. Scaffold planks. SANZ, Wellington.

Sulc, V. K. 1983. The grading of coconut palm sawn wood. Tech. Rep. 2 (draft only). Regional coconut wood training programme, Zamboanga, Philippines.

Suparman Karnasudirdja, Bakir Ginoga, and Osly Rachman. 1978. Strength classification of wood based on the relationship between modulus of rupture and other strength properties. Laporan Lembaga Penelitian Hasil Hutan (Forest Products Research Institute) 115, Indonesia (English translation).

Timber Research and Development Association. 1976. Grouping of Laos timbers for a community building system for Laos. Part 3 of a Report of the United Nations Industrial Development Organization, United Kingdom.

UNIDO. n.d. Wooden bridges. UNIDO's prefabricated modular system. Agro-Industries Branch, Division of Industrial Operations, United Nations Industrial Development Organization, Vienna.

——. 1985. Prefabricated modular wooden bridges. UNIDO/10/R.162. UNIDO, Vienna.

Wong, W. C., and C. N. Wong. 1980. Grouping species in plywood manufacture. Proceedings of Eleventh Commonwealth Forestry Conference, Trinidad.

Chapter 16

The New Composition or Composite Materials in Forest Products Trade

THOMAS M. MALONEY

Composition or composite wood products are an important part of the forest products industry, especially waferboard, oriented strandboard (OSB), laminated veneer lumber, parallel strand lumber, particleboard, and dry-process medium density fiberboard (MDF). Good export opportunities are apparent for these products, but it is also true that countries for which the trade is intended may use this new technology, based primarily on low quality raw material, to produce their own composition and composite products.

The major items of international trade in forest products over the years have been logs, chips, and lumber. In recent years, hardboard, softwood plywood, and hardwood plywood have become important items of trade. Now, particleboard, MDF, waferboard, and OSB, as well as small amounts of LVL (laminated veneer lumber), are on the international trade scene. The amount of LVL available for trade, and products made therefrom (trusses, I-beams), should increase greatly as this young industry grows.

For those involved in international trade, and for many in the domestic marketing field, the newer composition and composite products are familiar names, but what they are and how they compete in the building and nonstructural field is not clearly understood. The purpose of this paper is to describe these materials, what markets they now serve, and how they may affect today's markets as well as the markets of the future.

DEFINITIONS

A great deal of confusion on terminology exists not only worldwide but in the United States as well. The discussion on definitions below has been taken from a paper prepared for the IUFRO Eighteenth World Congress (Maloney 1986).

Currently the term "composite" is used to describe any wood material glued together, ranging from fiberboard to laminated beams and components. It would seem to be a simple matter to have names that are understood universally for the many wood and lignocellulosic composite materials. The descriptive problem is mostly in the area of composition materials. About thirty years ago, fairly well-agreed-upon terminology (such as fiberboard, hardboard, and particleboard) seemed to be developing to go along with the well-accepted term "plywood." Since then, however—for technical, marketing, personal, or unknown reasons—many more names for these products have come into existence. With this plethora of technical terms comes enormous confusion, not only for the uninformed but for those deeply involved in the industry.

This paper will (1) discuss the family of products or materials, concentrating on those made of particles and fiber, (2) review most of the terms used to date, and (3) and try to point out where the product names overlap and the confusion develops. A suggested general scheme for common nomenclature will be advanced, with full knowledge that many people will choose their own nomenclature no matter what is agreed upon by most of the industry.

HISTORY

The Family Name

The first problem encountered is what to call the *entire* family of products. "Composite" has already been suggested. I used the term "composition materials" earlier to cover the products made of fibers and various types of particles. This includes molded as well as panel products, thus ruling out the term "composition board." Some people prefer "reconstituted wood" as the proper term, but not all of the aforementioned materials are made of wood. Bagasse and flax are popular raw materials, as are other nonwood materials. Furthermore, the wood is still wood, not a material "reconstituted" into wood. Another popular term is "engineered panels or materials." Many of the products, however, are not engineered; they are simply made from available raw materials to fit a need.

In recent years, the word "composite" has become the popular term for composition materials, but a better use of this word is to describe all glued materials. "Composite," according to the dictionary, means made of "distinct parts." All of the products being discussed are wood- or lignocellulosic-based, but many materials are made of only one element (e.g., fiberboard). If there is any difference, it is in material geometry (e.g., veneer and flakes) not in the materials themselves. However, the term can be used as suggested, if everyone agrees. The problems come when describing subgroups such as particleboard.

To illustrate the confusion, the American Plywood Association (APA) defines a composite panel as one with veneer faces and a reconstituted wood core (APA 1983). One U.S. company includes only particleboard and medium density fiberboard in its composite division. Other panel products are in other divisions.

Until recently, the U.S. Forest Products Laboratory in Madison, Wisconsin included all panel products, including plywood, laminated veneer lumber, and parallel laminated veneer (as well as particleboard, flakeboard, waferboard, oriented strandboard, MDF, hardboard, fiberboard, and insulating board), in its research unit on structural composite products. Plywood processing is now in another division of the laboratory, but for organizational, not technical, reasons.

The problem of nomenclature was recognized long ago. A lengthy discussion on "Product Description, Nomenclature and Definitions" was held at the First International Consultation on Insulation Board, Hardboard, and Particleboard sponsored jointly by the Food and Agriculture Organization (FAO) of the United Nations and the Economic Commissions for Europe in 1957 (FAO 1958). In this discussion, the definition problem is noted and conflicting descriptions are given. Since that time, the problems in describing products have become worse rather than better.

Development of Names

Plywood will not be included in this discussion, although it is one of the most important wood-based panel materials. It is made of large sheets of veneer and its description should not be confusing to anyone. Also excluded are other glued products such as parallel laminated veneer and laminated beams, since there is little confusion about their description. They will, however, be discussed later when discussing the family of composites. The problems of confusion come with the many varieties of fiberboard, particleboard, waferboard, and oriented strandboard.

Fiberboard. Low density fiberboard (insulating type) was invented in 1914, as was mineral- or cement-bonded board (Maloney 1977). Low density fiberboard is specified in the United States as having a density of 0.16 to 0.50 g/cm^3 (AHA 1985). Usually mineral-bonded products are considered separately from other types of panels because mineral-bonded boards have inorganic binders while all others have organic binders. Since mineral-bonded materials have completely different characteristics, such as a very high density (well over 1.0 g/cm^3), from other wood-based materials, they will not be considered in this discussion.

Hardboard was invented in 1924 (Maloney 1977). In the United States, hardboard is specified as having a density of over 0.50 g/cm^3 (USDA 1973). Elsewhere in the world, medium density fiberboard (MDF) is made at densities between 0.40 and 0.80 g/cm^3 and high density fiberboard (hardboard) is made above 0.80 g/cm^3 in density.

The early definitions of fiberboard were associated exclusively with wet-process boards. With the development of dry-process methods of producing fiberboard, starting in 1947 (Maloney 1977), the previous density classifications were still appropriate. MDF (now the common name for dry-process medium density fiberboard developed in the mid-1960s) can be made at much greater thicknesses than the wet-process types. MDF also has different properties and, unlike most of the other fiberboard types, has furniture as its principal market. In the United States, MDF is usually made at densities between 0.64 and 0.80 g/cm^3, but there is no such specification in the standard (NPA 1980).

MDF is manufactured using a combination of particleboard and fiberboard technology, and at one time both the American Hardboard Association (AHA) and the National Particleboard Association (NPA) in the United States claimed MDF as a product covered by their respective associations. The American National Standard for "Medium Density Fiberboard for Interior Use" (NPA 1980) was agreed upon later by both associations, thus establishing type MDF as a new product. MDF was manufactured originally as a siding product and currently many companies are working on MDF for exterior uses. However, if MDF is made for exterior use in the future, a new element of confusion may arise.

Particleboard. Particleboard manufacture started in the 1930s (Maloney 1977) and has developed a wide range of descriptions, including chipboard, shavings board, and flakeboard. Some argue that waferboard and oriented strandboard are also particleboards. The terminology describing particleboard is agreed upon worldwide fairly well:

> [Particleboard is a] generic term for a panel manufactured from lignocellulose materials (usually wood), primarily in the form of discrete pieces or particles, as distinguished from fibers, combined with a synthetic resin or other suitable binder and bonded together under heat and pressure in a hot press by a process in which the entire interparticle bond is created by the added binder, and to which other materials may have been added during manufacture to improve certain properties. Particleboards are further defined by the method of pressing. When the pressure is applied in the direction perpendicular to the faces, as in a conventional multi-platen hot press, they are defined as flat-platen-pressed; and when the applied pressure is parallel to the faces, they are defined as extruded. (Quoted from ASTM D 1554, 1978)

Furthermore, there is agreement on most of the particle descriptions. According to the source quoted above, there are chips, curls, fibers, flakes, shavings, slivers, strands, and wood wool (excelsior). What is missing are two common particle types. The first is a "granule" which is found extensively in U.S. particleboard plants. This small particle is made by using hammermills or attrition mills and closely resembles a shaving. The second is the "wafer." The ASTM definition describes a flake as "a small wood particle of predetermined dimensions. . . ." The problem is, "How small is small?"

ASTM (American Society for Testing and Materials) has been addressing the problem of flake and wafer description and is discussing changing the definition of a flake. The definition of a flake would be changed to "a small wood particle of predetermined thickness . . . ," since many flaking machines control only the thickness.

ASTM is also considering changing the definition of strand to "a wood flake having a minimum predetermined length-to-width ratio of 2:1." This could be a particle less than 10 mm in length or as much as the 75 mm strands used in some plants.

The definition of "wafer" has been a problem. A wafer is a large flake and has been understood to be a relatively square particle usually about 40 mm long. The inventor of the waferboard process as well as the word "wafer," James d'A. Clark (1955), stated that a wafer has tapered ends. However, the wafers produced today are not necessarily made with tapered ends. In recent years, plants have been producing "oriented waferboard," which means that these wafers are more like strands. Clark also stated that wafers can be narrow in width. Thus there is considerable confusion over basic technology.

In addition, waferboards are made with wafers of many different sizes (within one waferboard) simply because many sizes are generated in the waferizing process. Most of the smaller wafers are placed in the board core. This can give the impression that a given waferboard has the same size wafers throughout the panel.

ASTM is also, as mentioned, addressing the problem of describing the wafer. For now, the proposed description calls for a wood flake having a predetermined length of at least 30 mm (some apparently want a description with a slightly longer length). There is no mention of width or whether the ends are to be tapered.

The Canadians have been struggling with the description of wafers and strands and definitions of the products made from them (CSA 1985). Wafers are defined as "a specific type of wood flake produced as a primary function of specialized equipment (i.e., waferizer) and having a controlled length of at least 30 mm along the grain direction, a controlled thickness, and a variable or controlled width. Each wafer is essentially flat and has the grain of the wood running predominantly in the plane of the wafer. In overall character wafers resemble small pieces of thin veneer. Wafers may be purposely produced with a narrow width to facilitate alignment." Strands are defined as "a specialized wafer having a length at least twice the width." Because strand, here, means a specialized wafer, the length has to be at least 30 mm long. Thus it really fits into the classification of a narrow wafer. Since the first great development of waferboard took place in Canada, it appears the name "wafer" is preferred over that of "strand" and the final product, which is OSB.

The ASTM D 1554 (1978) particle definitions were first developed in 1958. The two terms "fiberboard" and "particleboard," however, have not been used in all cases in the United States to cover these two groups of products according to its own ASTM definitions. Much of the rest of the world, however, has subscribed to this basic terminology.

In the United States, particleboard has come to mean boards made of planer shavings, granules, small flakes, and small fines (another undefined term). If such a board is used in house construction, it may be called structural particleboard (APA 1983). Most of the board, however, is bonded with urea-formaldehyde resin and is used for furniture, cabinets, floor underlayment, and mobile home decking. In the rest of the world, most particleboard is made with small flakes.

This brief recitation of the development of terminology was not meant to be a complete discussion, which would be a lengthy discourse, but rather to illustrate how confusion has developed, particularly in the United States, and to set the stage for describing the various materials.

SUGGESTED STANDARDIZED TERMINOLOGY

The position being taken is that all the materials under discussion are made up of elements that have been glued together. Their commonality is that they are composed of smaller pieces bonded or fused together. Thus in the forest products field, the term "composite" seems most appropriate, covering any material made of smaller pieces and glued together. Within this family of materials, several subgroups develop naturally. The differentiation between products within a subgroup could be on the basis of properties or uses rather than the type of particle or fiber used (e.g., roof sheathing, furniture panels, lumber) or by material description. This paper will concentrate on material description. The main subgroups of materials are listed in Table 1.

Table 1. The family of composite materials.

Veneer-Based Materials	Laminates
Plywood	Laminated beams
Parallel laminated veneer (PLV)	Overlayed materials
Usually made in wide billets	Panels or shaped materials
(e.g., 27 inches)	combined with nonwood
Laminated veneer lumber (LVL)	materials such as metal,
Cut from PLV	plastic, and fiberglass
Composition Materials	Edge-Glued Material
Fiberboard	Lumber panels
Insulation board	
Hardboard	Components
MDF	I-beams
Particleboard	T-beam panels
Waferboard	Stress-skin panels
Oriented waferboard (OWB)	
Oriented strandboard (OSB)	
COM-PLY®	
Parallel strand or aligned material	

The list in Table 1 is not all inclusive, but shows that a logical organization of composite materials is possible. Any organization of subgroups has to be flexible so that new materials developed in the future can be placed easily in the composite materials family.

THE MARKETS

As is well known, the forest resources used for manufacturing lumber and plywood are vastly different today compared with the past. The changes have been in the lower quality of logs harvested, the different species used, and the tremendous increases in energy costs that make it more expensive to ship products. These changes, along with new technology, have made it possible for waferboard and OSB to enter the structural market. Indeed, these factors have led to the development of composition (composite) lumber substitutes.

The main interest at present is in the area of structural panels, lumberlike materials, particleboard, and MDF. These products will be the subject of this discussion; but the special markets for components such as trusses, I-beams, and panelized construction which can be made more efficiently with these new materials should be recognized.

Structural Panels

For structural panels, softwood plywood remains the dominant product (about 22.1 billion square feet, 3/8 inch basis, in 1986) (R. G. Anderson, Director, Market Research and Economic Services, APA, Tacoma, Washington, pers. comm., 1987). However, nonveneer products have about 13% of the market (3.3 billion square feet, 3/8 inch basis) and are projected to have 25% by 1990 (Fuller 1985). On a strict percentage basis, this is impressive, particularly since most of the U.S. nonveneer production began during the 1980s. But this gain in production for nonveneer panels has not changed, to a great degree, the volume of plywood produced. The nonveneer production has contributed to the record production of structural panels over the past five years—a record not possible without the nonveneer plants in operation. It is projected by Fuller (1985) that the volume of plywood produced will start to decrease after 1990. Figure 1 shows changes in the production of veneer and nonveneer panels in the United States and Canada since 1952.

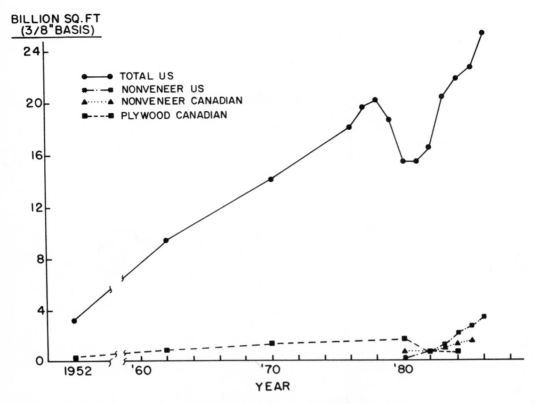

BILLION SQ.FT (3/8" BASIS)

Legend:
- ● TOTAL US
- ● NONVENEER US
- ▲ NONVENEER CANADIAN
- ■ PLYWOOD CANADIAN

Figure 1. Production of veneer and nonveneer panels in the United States and Canada.

A great advantage of nonveneer plants over plywood plants is that they do not rely on veneer-grade logs for raw material. They use lower quality logs, and great quantities of this type of raw material are found throughout much of the United States. Thus nonveneer plants are positioned strategically near major markets to reduce shipping costs (e.g., New England, the Lake States, the South). This gives them a great advantage over the plywood plants, particularly those in the Pacific Northwest, which are located near the source of veneer logs.

Also, nonveneer panels have an advantage in nonresidential construction, both here and abroad, because they can be used in sizes much larger than plywood. This gives them a great structural advantage. Large panels also make construction faster if machines are available to handle them, as they generally are in larger buildings under construction. Nonveneer panels in sizes such as 4 by 24 or 8 by 16 feet could be readily available. Yet to be worked out are engineering data showing that it is better to use such large panels than 4 by 8 foot panels, but in the writer's opinion the larger ones will be shown to be better in construction because of fewer joints.

Only small volumes of nonveneer panels are exported now. Users in other countries are just becoming familiar with them and have some questions about properties (thickness swell, for example). But as such problems are overcome through education and new technology, nonveneer panels will have far greater export opportunities.

A negative aspect is that countries with their own supply of low quality logs can build plants closer to markets than U.S. plants are. Whether this comes to pass on a large scale remains to be seen, but if it does, it will also affect the export of softwood plywood.

Nonveneer plants have already been built in Europe, and with success, more may be built. European plants do have access to species that fit into nonveneer panel production. This same situation is true in New Zealand and Australia. However, the mixture of hardwoods found in subtropical and tropical countries presents difficult manufacturing problems that could hinder non-veneer manufacturing development in these areas.

Lumberlike Products

Laminated veneer lumber (LVL) has been around for quite a while. Estimates of production today are about 13 million cubic feet or 104 million square feet (1-1/2 inch thick basis) per year. This material is used for flanges in I-beams, for truss parts, as scaffold planks, and for other uses directly competing with lumber. Examples are shown in Figure 2. It is not a product under development: it is in the marketplace and quite successful. It was used in the construction of the Summit House built last year in Japan to demonstrate the use of many types of wood materials in construction.

Figure 2. Examples of LVL (photograph courtesy of Trus Joist Corp.).

For the most part, the parallel laminated veneer (PLV) material manufactured for cutting into LVL is made of nondestructively tested veneers. The tested veneers are layered to produce the desired engineered properties in the final products. This manufacturing procedure would not be possible with a high degree of confidence without nondestructive testing. Thus this composite type of lumber is truly an engineered product.

An advantage of PLV is the production of the material in continuous presses. Thus materials can be cut to exact lengths. Furthermore, lengths of material for larger-than-conventional lumber are manufactured. Pieces wider than conventional lumber are also possible using PLV as the base material.

As with panels, LVL may be made elsewhere if suitable veneer material is available. Improvements in lathes may also make this possible in countries that do not now have veneer-quality logs (according to present grade standards). For example, some reported Japanese developments in slicing veneer from smaller logs may allow the use of logs now considered too low in quality to use for veneer.

Developments in producing lumberlike products of strands or as COM-PLY® are also on the scene. MacMillan Bloedel has a pilot plant making parallel strand lumber (PSL), a high quality material available in long lengths (up to 80 feet) (Figure 3). This material was used in constructing many of the buildings for the 1986 World's Fair in Vancouver, British Columbia, Canada. A plant

Figure 3. Examples of PSL (photograph courtesy of Parallam Division of MacMillan Bloedel Limited).

producing COM-PLY® lumber is now starting operation in North Carolina. Figure 4 illustrates the makeup of COM-PLY® lumber.

Some companies are considering producing aligned flakeboard type material for use as lumber. This has been shown to be possible by numerous research studies. The potential of such material-was well illustrated in a Washington State University study (Maloney 1984).

In considering the actual design values, an example of some recent work on oriented strandboard made from nominal 1.5 inch long strands could be considered. The research showed that such a product could justify a higher allowable tensile stress than is assigned currently to visually graded

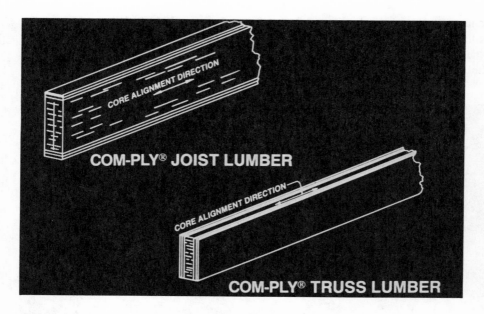

COM-PLY® JOIST LUMBER

CORE ALIGNMENT DIRECTION

CORE ALIGNMENT DIRECTION

COM-PLY® TRUSS LUMBER

Figure 4. Schematic drawing of two types of COM-PLY® lumber.

Select Structural Douglas-fir lumber. This research was done on oriented flake materials intended to substitute for 2 by 4 sawn truss lumber. In this material the particles are aligned in one direction to simulate sawn lumber. Thus this could be called oriented strand lumber (OSL).

Testing of visually graded 2 by 4 by 8 foot long Douglas-fir Select Structural lumber (113 specimens) showed a range of tension values from 1,280 to 10,250 psi with a mean of 7,250 psi. According to the Western Wood Products Grading Rules (WWPA 1980, p. 202), the allowable working stress in tension for this product is 1,400 psi. Using the ASTM method D 2915 (1982) for determining the allowable working stresses for visually graded material, the targeted test value for any of the material is 2,930 psi. Thus some of the 113 boards were below the test value required to justify the allowable tension value for select structural lumber. Most of the material, however, was far above the minimum value and consequently was much better than needed—a waste of quality material.

The OSL material had tensile strengths ranging from 3,260 to 4,000 psi (a very tight range or much better coefficient of variation compared to the sawn lumber). All values were well above the minimum design value of 1,400 psi. This example shows clearly how a low grade raw material can be converted into a high grade structural material exceeding the properties of the best grade of structural lumber available. It is easy to make the argument for nondestructively testing Select Structural Douglas-fir so that the majority of the lumber can be used for higher value applications. This obviously should be done. However, this example shows also, with the strength property discussed, that OSL can be used for all of the common uses of high strength fir lumber—an exciting possibility.

As with LVL, it is possible to manufacture OSL from wood raw materials elsewhere in the world. Quality sawlogs or veneer logs are not necessary. Technology can overcome the lack of quality roundwood. The United States, unlike much of the rest of the world, has excellent species for producing these new lumberlike materials out of small particles. Higher density hardwood species, to date, are not as suitable for raw material as the U.S. softwood species and low density hardwood species. There is always the possibility, however, that technology will solve the problems of using these more difficult-to-use species.

Particleboard and MDF

Particleboard and MDF are being considered together because their major markets are furniture and cabinet manufacture. Some exporting has been taking place, but to the writer's knowledge there are no official figures. It is known that special grades have been produced for some specific overseas customers.

Particleboard and MDF are produced from low grade wood raw materials (mainly planer shavings) in the United States. Roundwood, unacceptable for most forest products manufacture, is quite suitable for particleboard and MDF manufacture. In fact, other lignocellulose materials, such as flax and bagasse (sugar cane residue), are used.

Thus there are suitable raw materials worldwide for making particleboard and MDF. The advantage the United States has, at least for the present, is the quality of the species being used. Figure 5 shows the production of particleboard and MDF in the United States (particleboard since 1961 and MDF since 1975). Both are major product lines, a fact often overlooked by those interested mainly in lumber and plywood.

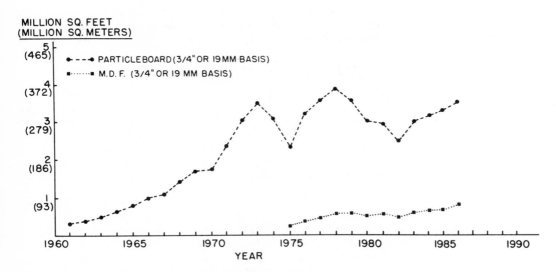

Figure 5. Production of particleboard and MDF in the United States.

Some Asian countries have developed sophisticated plants using the latest technology for converting raw particleboard and MDF into finished furniture parts (laminated with plastics, edge finished, etc.). These parts are shipped to the United States ready for assembly. This is making an impact on the U.S. furniture industry, reducing its manufacturing capability (Fuller 1986). Particleboard and MDF are being shipped from the United States for this use; however, if board-producing plants are built in Asia to supply this market, both the U.S. domestic and international markets will be affected adversely.

CONCLUSION

The composition/composite materials sector of the forest products industry is well advanced and important in its own right, both domestically and internationally. Technological advances are coming about rapidly to enhance the development of the industry. It is a powerful competitor to the lumber and plywood sectors and probably will become even more so in the future.

The technology developed can also be used elsewhere. For example, the technology could be used where poor quality wood or other lignocellulosic materials can be converted into building products. New plants located in other parts of the world could become quite competitive with U.S. plants if they are close to the targeted markets.

REFERENCES

AHA. 1985. American national standard—Cellulosic fiberboard. ANSI/AHA A194.1. American Hardboard Association, Palatine, Illinois.

APA. 1983. Performance rated panels. American Plywood Association, Tacoma, Washington.

ASTM. 1978. Standard definitions of terms relating to wood-base fiber and particle panel materials. ASTM D 1554. American Society for Testing and Materials, Philadelphia.

——. 1982. Standard Method for evaluating allowable properties for grades of structural lumber. ASTM D 2915. American Society for Testing and Materials, Philadelphia.

CSA. 1985. Waferboard and strandboard. National Standards of Canada CAN3-0437.0-1785. Canadian Standards Association, Rexdale, Ontario.

Clark, J. d'A. 1955. A new dry process multi-ply board. For. Prod. J. 5(4):209-213.

FAO. 1958. Fiberboard and particleboard, pp. 4-14. Report of an International Consultation on Insulation Board, Hardboard, and Particleboard, Geneva, Switzerland. Food and Agriculture Organization, Rome.

Fuller, B. 1985. Economics of substitution: Projects for nonveneered and veneered structural panels and other nontraditional wood products. In T. M. Maloney (ed.) Proceedings, Nineteenth International Particleboard/Composite Materials Symposium, pp. 21-51. Washington State University, Pullman.

——. 1986. Factors affecting demand and supply for composite board products. Paper presented at Forest Products Research Society's Composite Board Products for Furniture and Cabinets—Innovations in Manufacture and Utilization Conference, Greensboro, North Carolina.

Maloney, T. M. 1977. Modern particleboard and dry-process fiberboard manufacturing, pp. 31-39, 179-181. Miller Freeman Publications, San Francisco.

——. 1984. Electrically-oriented composition boards made with smaller size strands and particles. In Proceedings of Conference on Comminution of Wood and Bark, Forest Products Research Society, Madison, Wisconsin, pp 86-95.

——. 1986. Terminology and product definitions: A suggested approach to uniformity worldwide. In International Union of Forest Research Organizations (IUFRO) XVIII World Congress Proceedings, Division 5, Forest Products, Ljubljana, Yugoslavia, pp. 294-305. Obtain copies from IUFRO Secretariat, Vienna, Austria.

NPA. 1980. American national standard medium density fiberboard for interior use. ANSI A208.2-1980. National Particleboard Association, Gaithersburg, Maryland.

USDA, National Bureau of Standards. 1973. Voluntary product standard PS 58-73. Basic hardboard. U.S. Department of Commerce, National Bureau of Standards, Washington, D.C.

WWPA. 1980. Standard grading rules for western lumber. Western Wood Products Association, Portland, Oregon.

Chapter 17

Nonresidential Wood Construction in Japan and Japanese Building Codes

YUJI NOGA

HISTORY OF WOOD USE IN JAPAN

Wood has long been a basic building material in Japan. Methods of wooden construction originally developed in China have been in common use in Japan for many centuries. Not only were the famous temples and shrines constructed of wood, but the majority of other structures as well, such as residential, commercial, and public buildings. The basic method devised is similar to what is known as post and beam construction. Wooden construction was highly developed and intricate in detail. However, the use of wood slowly began to recede during Japan's era of modernization beginning with the Meiji period. It was during this period that Japan established its first building code adapted from Western countries.

In terms of its ultimate effect on wooden construction, probably the most significant aspect of Japan's original building code was the attempt to control the hazards of fire by isolating the fire within one structure. By restricting the use of combustible materials on a building's exterior, a fire would not so easily spread from one building to another. This basic philosophy of restricting the use of combustible materials within fire zones continues to strongly influence Japan's building code today. Other Western countries have moved away from this approach by recognizing the efficiency of limiting and isolating fire hazards on a room-to-room basis. This is done by fireproofing combustible materials and limiting the spread of fire by constructing according to a fire rating system.

Japan's building code continues to exclude wood in any form in the definitions of "fireproof construction" and "noncombustible materials." This effectively prohibits the use of structural wooden construction within any fire zone or in medium scale public and commercial buildings.

The fire zoning way of thinking was reinforced in Japan through its experiences with the great Kanto earthquake in 1926 and the fire bombings of World War II. Wooden construction suffered even more in the hasty rebuilding efforts, because poorly prepared materials and short-cut construction techniques were used. The result was a damaged image of wood construction. Although the traditional appreciation for the aesthetic qualities of wood was maintained, the integrity of the performance of wood was lost.

After World War II, the building codes became more severe in restricting the use of wooden construction methods. An almost totally deforested Japan turned strongly toward more readily available materials such as concrete and steel. The continuing poor performance of hastily rebuilt structures solidified the backward image of wooden construction, while at the same time concrete and steel represented Japan's technological advancement.

Wooden construction has continued to decline in postwar Japan to the point where today only 46% of residential construction uses wood structures; many temples and shrines are rebuilt with reinforced concrete; and wooden construction is virtually excluded from public and commercial buildings.

This declining use of wood has recently been noticed by industry and government concerns. Efforts are being made to promote the use of wood and to rekindle Japan's traditional cultural affinity for wood. An abundant supply of wood, both domestic and foreign, also is a motivating factor. In addition, technological advancements in production methods and in engineered wood products contradict many of the reasons that led to the restrictive building codes and ultimately to the decline of wooden construction.

REVISION OF BUILDING CODES

One of the basic necessities to facilitate and promote the use of wood in Japan is a revision of the building code. A committee has been appointed (Table 1) to formulate recommendations to the Ministry of Construction after a five year investigative period (1986-90). Throughout this period various aspects of wood frame construction techniques will be reviewed.

Table 1. Committee for Development of New Timber Construction Technology.

Chairmen

Professor Yoshita Uchida, Meiji University
Professor Hideo Sugiyama, Tokyo Science University
Professor Kachi Kishitani, Tokyo University
Professor Masata Yasuoka, Tokyo University

Architects

Yuji Noga, Issiki Architects & Partners
Masao Hayakawa, Hayakawa Architectural Design Office
Masaya Fujimoto, Gendaikeikaku Kenkyusho

Foreign Associations

Rick Skorick, American Plywood Association
Joseph Caron, Council of Forest Industries of British Columbia

Industry and Professional Associations and Unions

Architectural Center of Japan
Japan Housing and Wood Technology Center
Japan Architect Association
Contractor's Association
Japan Contractor's Union
Japan Timber House Industry Association
Japan 2 x 4 Homebuilder's Association
Prefabricated Housing Association
Japan Federation of Lumber Association
Japan Laminated Lumber Manufacturer's Association

Government Organizations

Japan Housing Loan Corporation
Japan Housing Corporation
Ministry of Education
Forestry Agency
Department of Fire Prevention
Ministry of Construction
Building Research Institute
Land Development Technical Research Center

The committee members reflect the various interests regarding a revision of the building code. Tokyo and Waseda universities are well represented by top academicians involved in timber design, construction, and research. Private architects with extensive experience in wood construction design are also participants. Representatives from two major foreign interests—the Council of Forest Industries of British Columbia and the American Plywood Association—sit on the committee. Also included are Japanese industry associations and labor union representatives. The Japanese government is well represented by the Japan Housing Loan Corporation, the Japan Housing Corporation, and other related ministries and departments. The committee's agenda and activities are basically managed by its secretary general, Hideo Totani of the Land Development Technical Research Center. Mr. Totani was instrumental in the effort to introduce the North American platform frame construction method in Japan over ten years ago.

The committee will concern itself with three general areas for review (Figure 1): (1) it will look into technological developments concerning structural timber engineering; (2) it will conduct a technical review of fire prevention planning; and (3) it will review actual living environment and structural durability data.

Under the topic of structural timber engineering technology, the committee will review various material grading systems, standardization of structural performance specifications by wood species, standardization of structural engineering specifications, the use of hardware connectors, and the

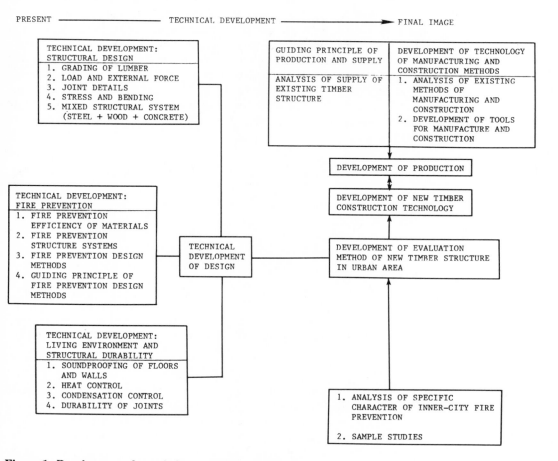

Figure 1. Development of new timber construction technology.

integration of different structural systems within the same structures. The number of different construction systems that have been developed by the over 300,000 construction companies and home builders of Japan has resulted in systems that lack standard grades and specifications.

Probably the most important topics to be reviewed are those related to the technical review of fire prevention planning. A lot is at stake for the lumber industry and construction businesses. Specific topics of concern include the efficiency of various fireproofing and fire retarding materials, a comparative analysis of the different construction methods, the effect of design, and urban planning concerns.

Finally, aspects that relate to the living environment and structural durability will be reviewed. This includes soundproofing, heating-cooling insulation, moisture condensation control, and the durability of the various structural components and connectors. Japan is intent on increasing both the quality of the living environment and the durability of its structures.

FUTURE ISSUES RELATED TO WOOD FRAME CONSTRUCTION

In today's marketplace in Japan there are at least five issues that could significantly affect the future level of wood frame construction. The committee will have to grapple with these issues throughout the course of its review.

The key issue is whether Japan will incorporate fire rated wood structures in its building code definitions of "fireproof construction" and "noncombustible materials." As stated earlier, Japan's basic approach has been to isolate fire from structure to structure by prohibiting the exterior use of flammable materials. This fire zoning approach has been abandoned in North America in favor of a fire rating system that attempts to isolate and contain the fire within a single room long enough to be extinguished. A 1982 comparative analysis conducted by Dr. Kobaynishi revealed that 30.5 deaths result per 1,000 occurrences of fire in Japan; the figure for the United States for the same number of occurrences is 2.4 deaths (Table 2). This type of data would suggest that the fire rating approach has great merit, and specifically that the exclusion of wood frame construction does not ensure a safer environment.

Table 2. Fire occurrences and death rate by country, 1982.

Country	No. of Fires	Deaths	Deaths/1,000 Occurrences
Japan	60,568	1,849	30.5
United States	2,588,000	6,137	2.4
Canada	76,199	675	8.9
England	357,924	919	2.6
West Germany	142,000	1,000	7.0
South Korea	6,822	276	40.5

Source: H. Totani, "Fire Zoning in the U.S.A.," in Tochi Tutaku Mondai (ed.), *Land and Housing Problems*, vol. 2 (1987).

The Japanese government has shown strong interest in developing an acceptable fire rated wood frame system. It has conducted a number of full-scale burn tests in Japan, the most recent one in January 1987. In that test two houses were built and burned, one made with fire rated materials and the other without. Fires could be contained twice as effectively in the fire rated house. Design factors such as window openings were also studied. All this activity generates the necessary data the Japanese committee will need to reach the objective of broadening the restrictive definitions in the building code to include and allow fire rated wood frame structures.

Another important issue the committee will face is closely intertwined with changing the fire code: the acceptance of three-story wood frame construction. In Japan, one of the biggest growth markets is three- and four-story inner-city structures needed to replace the hastily built postwar structures. The object of the committee here will be similar to the previous issue: redefine "fireproof construction" and "noncombustible materials" to include fire rated wood frame construction.

A third issue is related to the use of green lumber versus dried lumber. Traditionally, Japanese wood frame construction used only carefully selected, air dried lumber. But because of the great demand for lumber following the war, as a result of fire bombing and then the sudden urban growth, there was no opportunity to take such care in selecting lumber. Green wood of various species was used in structures within one month of harvest. The resulting structural performance and durability problems are largely responsible for the public's poor image of wood frame construction. The use of kiln dried wood can help reestablish the integrity of wood frame construction.

Another issue is the exterior use of wood products. Japanese are not accustomed to using wood for fences, decks, siding, and so forth. The attitude is that wood rots and is not suited for the hot, moist Japanese summer climate. There has been a strong rejection of the exterior use of organic materials such as wood. However, the proper use of chemically treated wood in this difficult area could not only open a new market but help regain the integrity of wood in the mind of the public. Once again, the object must be to change the fire code, because it now requires any exterior wood application to be completely covered with stucco—a requirement that only contributes to the deterioration process.

A final issue is one that until recently was totally overlooked: the use of wood in civil engineering projects. This market is potentially very large in Japan, and it is surprising that it has been ignored in trade negotiations. The aesthetical value of wood in such uses can add greatly to the final product. A minor but telling point is that the literal translation of the Japanese term "civil engineer" is "one who builds with earth and trees." Civil engineering in Japan today is the exclusive realm of concrete and steel.

In conclusion, although Japan and the Japanese clearly have a high appreciation of wood and a long heritage in the use of wood, the building code stands in the way of allowing wood frame construction to flourish in Japan.

Chapter 18

Nonresidential Wood Structures in the United States

MARSHALL TURNER

Wood is an excellent building material. It is strong yet light in weight; it is easy to work; it absorbs dynamic loads; and it has an inherent warmth and beauty possessed by no other structural material. Wood has a definite place in the nonresidential construction market, yet it has not come close to penetrating this market.

By tradition, wood used in construction in the United States is restricted to residential buildings. The nonresidential market is dominated by steel and concrete. This state of affairs did not happen by accident. Builders of steel structures have developed a very competitive approach. Basically, they control all stages of the structure's evolution by providing a complete "turnkey" package. This is their strength—one that should be kept in mind when considering the nonresidential market.

There have been developments, however, that make it possible for wood products to compete in the nonresidential market with the dominant materials. Let us briefly explore some of the historical events that have established the excellent structural capability of wood.

Undoubtedly the first bridges were logs across rivers. Some of the earliest shelters were constructed with wood poles, with sticks forming the small supporting elements. The shaping and molding of wood for building systems occurred slowly, evolving over many centuries. China and Japan, for example, had well-developed timber framing systems dating back more than 2,000 years. Elaborate timber building parts were standardized, premanufactured, and actually listed in catalogs. If the emperor needed a certain size royal building for a particular site, the appropriate pieces would be selected and sent to the location, there to be assembled.

For these early structures, the builders took advantage of the excellent compression properties of wood. Mine supports are good examples of structural timber members that utilize this property. Signs of distress show up early in these supports, but the residual compressive strength allows for a certain margin of safety.

Wood became efficient as a long-span engineering material when economical bolted or glued joints were developed to accommodate tension forces. These developments originated in northern Europe partly because Germany was short of steel during the war years of this century, and needed design improvements to make use of the country's small-diameter wood resource. This led to sophisticated connection systems such as split rings, shear plates, and the like. Structural finger

Editor's Note: Two slide programs were shown as part of Mr. Turner's presentation. The first, produced by the Western Wood Products Association (WWPA), dealt with wood construction in the nonresidential market. The program is available from the WWPA upon request. The second was produced by Western Wood Structures, a firm owned by the speaker. This firm, in addition to building large nonresidential wood structures and timber bridges, provided the dome for the Tacoma Dome. At the time of this symposium, this 530-foot-diameter municipal structure is the largest wood dome building in existence. It is used successfully for a variety of sporting, entertainment, and exhibition events. This paper is a synopsis of the speaker's oral presentation. A few photographs from the slide presentations are included.

joints were developed as waterproof synthetic glues replaced natural glues. The use of these connectors led to the efficiency of wood use for structural applications.

Finally, the introduction of computers has allowed wood to compete with long-span steel. Recently developed design programs allow an engineer to analyze alternative designs and loading conditions very rapidly. Because the best configuration for the lowest cost can be found, wood can be used in the most efficient and reliable manner.

Given this background, the question yet to be answered is: How can we increase market share of wood in the nonresidential market throughout the world? Before addressing this question, a brief description of the nonresidential market in the United States is needed.

NONRESIDENTIAL MARKET IN THE UNITED STATES

The nonresidential market in the United States is huge, estimated at $75 billion a year (Dodge/Sweet 1987). In the eleven western states where timber is accepted, this market is near $19 billion, leaving a $56 billion market virtually untapped. About 85% of this construction market is one- to three-story buildings, in five categories: schools, warehouses, churches, office buildings, and stores. Of these, stores account for 25 to 30%.

One- to three-story buildings are ideal for heavy timber construction. Aside from the steel and glass skyscrapers, wood can penetrate the five categories, domestically and internationally.

CHANGES NEEDED

A three-phase program must be initiated before wood can effectively compete in the world nonresidential market: (1) An educational program must be developed for and aimed at the construction team (building owner, developers, and managers) as well as architects and engineerings. (2) A concentrated effort must be made to remove unfair building codes and insurance barriers. (3) A promotional effort must be initiated that is designed to make the building team comfortable with using wood.

It is ironic that the same universities that have provided excellent training to large numbers of wood technologists (while increasing our knowledge of timber properties) are the ones whose schools of architecture and engineering provide much training in steel and concrete design—but very little in wood.

According to Dr. James Goodman, Colorado State University, a critical need exists for a worldwide education effort in wood engineering. He made this statement during the 1983 Workshop on Structural Wood Research (Goodman 1983). He also said that those engineers involved in the design of structures must be better educated in the unique behavior of wood, and in the application of this knowledge to the design procedure. Clearly, this education gap must be filled.

It follows that building codes must also be changed to recognize new developments in timber engineering. Oregon, for example, a state with vast reserves of the best structural wood in the world, has a highway department that forbids the construction of wooden bridges. Numerous other examples can be cited where timber could be economically and effectively utilized in construction, but well-established, unfair codes prevent wood use. This change will not occur easily or quickly. It requires a sustained effort—one that is concentrated, coordinated, and even cooperative.

In the same vein, insurance barriers must be reconsidered. Wood, in fact, has two peculiarities that seem to cloud the issue of its reliability: (1) it burns well when broken into small, dry pieces; and (2) it rots if kept wet and damp. It should be apparent that large pieces of wood do char during a fire, but take a long time to "burn" enough to reduce the load-carrying capacity of the structure.

Perhaps an example to illustrate this point is worth mentioning. The Tacoma Dome has experienced one significant fire since it was built. The events leading to the ignition of the fire and its being extinguished began on a Thursday night, when 30,000 people filled the Dome to hear the rock group AC/DC. As part of the show, the house lights were turned out while pistols with blank ammunition were shot off on stage. During this blackout, an on-leave soldier in attendance fired a battlefield flare gun. The hot projectile stuck in the 2 inch wood roof decking. The soldier was arrested (and subsequently court-martialed), and the episode seemed finished.

Figure 1. The Tacoma Dome during construction. At 530 feet in diameter, this is the largest wood-domed structure in the world. Major members and intermediate purlins are Douglas-fir glued-laminated timber; decking is nominal 2 inch solid sawn Douglas-fir. The Dome is owned by the city of Tacoma, Washington, and is used for major sporting, convention, and entertainment events.

Three high school football games were played in the Dome on Friday, and it was not until 10 o'clock Saturday morning that someone in the vicinity noticed smoke coming from the roof. The fire truck arrived and quickly extinguished the smoldering area, but water pressure blew a 32 square-foot hole in the roof. This was the worst damage. No structural damage had occurred after a 9 inch shell, with a burning temperature estimated near 1400°F, smoldered in the timber roof for thirty-six hours. The news media heard about the incident and arrived around 2 o'clock Saturday afternoon to find no traces of damage. The building roof had been closed, because another concert was scheduled for Sunday evening, with 25,000 seats already sold. The needed repairs, including new roofing, had been made within a three-hour period.

Yet despite the demonstrated safety of a wood dome, the city of Tacoma was obliged to cancel fire insurance coverage on the arena because of escalating premium costs. The city was able to cope by placing a fire truck at a hydrant in the structure during events that represented a fire danger. The existing television security monitors were used to scan for evidence of fire. Subsequently, the insurance coverage was reinstated, but only after the insurers did a bit of homework to arrive at new rates.

Figure 2. Worker installing a Glulam beam/wood I-joist floor system on a three-story commercial structure.

Figure 3. This bridge uses Glulam bowstring trusses, timber decking, and timber handrails to span 170 feet. The bridge carries pedestrians as well as municipal water and sewer lines.

This example indicates that properly designed heavy timber structures can be as safe as nonwood structures, or safer. Generally, insurance rates do not reflect this truth.

Promotion is another area that needs a lot of attention. The Western Wood Products Association (WWPA) has recently produced an attractive slide program that deals with the nonresidential market. This kind of effort is needed, but much additional work remains to be done.

SUMMARY

I believe that wood will continue to prove itself as second to none when it comes to creating the built environment, if we are able to form strategic partnerships with the owners, architects, engineers, and builders of schools, churches, warehouses, office buildings, shopping centers, and the like. The success of wood depends on the person who specifies the building materials that contribute to the safety, economy, and beauty of these environments.

We cannot claim that wood is always the best or most appropriate material. We know that technology can find a substitute to perform any one of wood's sturctural functions better, and often cheaper. But there are qualities about wood, in both aesthetic and design potential, that can contribute to enriched living in ways that the sometimes narrow aims of technology fail to recognize.

REFERENCES

Dodge/Sweet. 1987. Construction outlook. McGraw-Hill Information Group.

Goodman, R. 1983. Reliability-based design for wood structures. *In* Proceedings, Workshop on Structural Wood Research, ASCE, Madison, Wisconsin.

Tropical Countries as
Suppliers and Consumers

Chapter 19

Tropical Countries: Suppliers or Consumers of Forest Products?

JAMES S. BETHEL

The countries that occupy the world's tropical zone, in whole or in part, vary in size and in the fraction of their areas dedicated to forest use. Brazil, for example, has a total land area of about 846 million hectares, of which more than two-thirds is forest land. On the other hand, a small country like El Salvador has a land area of just 2 million hectares, and less than 7% of this is in forest. A few tropical countries still have large areas of primary natural forest while others have essentially none. The norm, if there is one, is for a well-forested tropical country to be substantially supplied with secondary forest that has resulted from repeated high grading of the original natural forest.

In some of these tropical countries, forests represent a most important land use. Of all the tropical countries of the world, 26% have more than 50% of their land area in forest. For others, notably those located in the arid tropics, forests are not a significant natural resource. The countries that include tropical forests within their boundaries also vary in size, culture, and political character. In the case of some large countries, only part of the forest resource is in the tropical region. Brazil is an example of such a country. This variation among forests in the tropics has a profound influence on the contribution that a particular forest can make to materials supply. A country that has a large and varied forest resource may be a substantial exporter of forest commodities or forest products. A country with a very small forest resource may be a potential customer for exported forest commodities or forest products.

The tropical region as a whole is made up primarily, but not exclusively, of poor countries (i.e., countries whose per capita income is less than U.S.$500). Nonetheless, these mostly poor countries, in the aggregate, represent a huge international market for forest products. Some of them are very large suppliers or potential suppliers of forest products to world markets. The purpose of this paper is to suggest criteria for evaluating a tropical country as a trading partner with the United States.

TROPICAL COUNTRIES AS SUPPLIERS

From a technical standpoint, a country that aspires to be an exporter of wood must have a good raw material supply, an ample energy supply, a pool of technical manpower, and access to capital. If all of these resources are available, a country ought to be able to develop a significant forest products manufacturing and marketing program.

Raw Material Supply

A characteristic of many natural tropical forests is that they may be very poor as sources of merchantable materials. Typically they are all-aged, multispecies hardwood forests. Often only a few of the species are known in the export market.

Secondary natural forests that we have studied in Asia and Latin America may have a standing tree volume of 200 to 300 cubic meters per hectare but yield only 5 to 10% of that in marketable rough lumber or plywood. This may result in an expensive product even though the price of stumpage is very low. If fuelwood is added to the product mix, the yield can be considerably greater and the price of the lumber and plywood components will be reduced. Similarly, if the domestic market can absorb the lower grades of product, the upper grades can often be exported at a profit.

One of the reasons that these forests are poor as sources of raw materials is that many tropical countries have not developed good domestic markets for their own forest products from material that cannot be exported. The price of exportable commodities and products can be kept competitive if the lower product grades and the unmerchantable export species can be sold in local markets.

It is much easier to develop an export market for a species that is unknown in world trade if it can spin off a domestic use and a solid domestic market. The products that are relatively indiscriminate with respect to species, size, and qualities generally cannot command the prices that permit transport to remote export markets. Similarly, the lower grades of lumber or plywood cannot carry heavy transport charges and remain competitive. Accordingly, the opportunity to serve a substantial and diverse domestic market expands a country's export potential.

If the sale of forest products in world trade is based on the use of natural forests, either primary or secondary, the source of raw material is likely to be a declining one. Secondary forests that follow the first harvest may contain few of the original export species. Some natural forests, of which the dipterocarp forests of Malaysia and Indonesia are examples, are rich in a few species that are merchantable in export markets and therefore are exceptions to this rule. These forests can provide significant yields of products such as lumber and plywood from secondary forests. There are a few other tropical forests that are essentially one-species forests and accordingly are rich in resources, such as the mangrove forests of Africa, Asia, and Latin America and the cativo forests of the Caribbean coastal zones. But most of the forests of the tropical regions are low in product yield. Many of them would be considered noncommercial using any rational industrial criteria.

Man-made production forests are likely to be less diverse than natural forests. Often they are made up of exotic species. Recovery of product is usually higher than for natural forests by several orders of magnitude. Yield per hectare per year is also typically much higher. The tropical countries have the potential for developing man-made production forests that are very rich in exportable materials. These forests can be, and usually are, dependable long-term materials supply resources. At this time there are very limited plantation based forests—about one-half of 1% of the area of natural forests with trees. But the area of these forests is increasing very rapidly.

Energy Supply

The energy resources generally available in tropical countries that can be used to produce forest products for export are fossil fuels, hydroelectric power, animal power, manpower, and fuelwood. Some tropical countries have indigenous supplies of fossil fuels, such as Mexico, Indonesia, and Nigeria. Countries that do not have developed fossil fuel resources typically find that these sources of energy are very expensive when they have to be imported. A significant number of tropical countries have hydroelectric potential, but often it is undeveloped. Most tropical countries that are able to export forest products have substantial fuelwood resources. Animals are important sources of energy in some tropical countries. Elephants are extensively used for logging in Thailand and Burma. Manpower is available as an energy resource in most tropical countries.

Technical Manpower

Manpower can contribute to the development of an export program in forest products in tropical countries in two important ways. We have already seen that manpower can be a source of energy to be used in logging and milling. A second and more important manpower role is that of decision making (i.e., the use of the intellect in choosing actions among alternatives). Professional and technical education and training are prerequisites to undertaking technological roles in the production process. The need for trained personnel varies from product to product and from operation to operation.

Logging may require a relatively small fraction of employees with important decision-making capability and a much larger fraction whose principal role is to contribute energy to the process. This may not be true if a very sophisticated logging system is to be employed.

In the field of pulp and paper manufacture the judgmental role of the workman is much more important than the energy supply role. Manufacture of such products as lumber, plywood, particleboard, and fiberboard typically use manpower for both energy and technical skills. These uses are often combined in the same task.

If the pool of technical manpower in a tropical country is too small to support a forest products export program, it must be augmented if such a program is in the national interest. This may mean that the essential technical manpower requirements must be provided by expatriates, at least long enough to permit the country to overcome its technical manpower deficiency.

Access to Capital

Even when a tropical country has a significant raw material supply, it may not be able to provide a dependable supply of exportable forest products unless it has access to capital. Poor countries have several potential sources of capital that can be accessed for the development of a forest products export program.

A major source of capital may be the standing merchantable volume of timber in the primary forest. This was a major source in the early development of a U.S. export capability. It has been used effectively by some tropical countries such as the Philippines, Malaysia, and Indonesia. Other sources are capital input to projects from external trading partners or joint venture partners. Still others are equity investment and access to financial markets both domestic and foreign. This includes access to the World Bank, regional development banks, and international and binational technical assistance programs.

It is important for importers or potential importers who are interested in developing a dependable source of supply of tropical forest products to examine these elements of the resource base of the country of interest. Where there is a favorable level of all these industrial resources, the country should be able to develop a sound forest products export base provided that the political and institutional climate is favorable to world trade. A critical examination of these resource elements is particularly important if the foreign corporation is expected to provide capital, management capability, marketing service, or technology transfer as a condition of engaging in the trade enterprise.

TROPICAL COUNTRIES AS CONSUMERS

When U.S. forest products firms undertake to enter the international markets for wood, they generally look to the established world markets. These are generally in the developed countries of Europe, Japan, Canada, Australia, New Zealand, and the like. Certainly these are the markets where there are buyers knowledgeable concerning U.S. species and product specifications. This is perhaps the easiest entry into international trade, but it is not without its problems. Competition in these markets can be fierce not only with other foreign suppliers but often with domestic producers whose position in the market is often protected by trade barriers of all kinds.

The tropical countries, on the other hand, offer a different set of opportunities. In total, the tropical countries of the world represented about 15% of world wood imports in 1984. These imports were distributed among 117 countries that varied widely in size, biogeography, political organization, and culture. The share of the total imports for these tropical countries varied over a very broad range from essentially zero to just under $900 million in 1984.

The tropical countries as a group are growing much faster as markets for forest products in world trade than are the well-established markets of the developed industrial countries. Some of the countries that import the largest amounts, Egypt and Saudi Arabia for example, have very limited forest resources of their own. But even among the tropical countries that have substantial forest resources and that are themselves the exporters of forest products, there are good opportunities to develop export markets. This is partly because tropical forest resources are quite different from temperate forest resources and many of the tropical countries have demand for products that cannot be produced from indigenous sources.

The U.S. firm that wishes to develop long-term export markets in countries with rapidly growing demand for forest products must be prepared to select its customer country with care and must also be prepared to gain a substantial understanding of that country. Among the questions that need to be asked when contemplating development of an export market in a tropical country are: (1) How is wood typically and traditionally used? (2) In what forms are forest products required? (3) What business practices are customary in importation? (4) What language is usually used as the basis for international trade? (5) What is the competition? Who is currently supplying the market? (6) What port facilities are available to handle imported bulk cargo? (7) What transport facilities are available to move the cargo from the country of origin to the appropriate U.S. port?

It is important for the firm contemplating the long-term development of a market in a tropical country to consider the factors likely to influence the growth of the market and its accessibility to a U.S. company.

Political Climate

Some of the tropical countries that are good prospects as customers for forest products may not be accessible for international political reasons. Iran, for example, was a major importer of forest products in 1984. Given the current state of international politics, this is not a good candidate to be a trading partner of a U.S. firm. On a much more subtle level, there are other tropical countries where a U.S. company would be at a competitive disadvantage in developing an accessible market. Often favored-nation status is shared among countries formerly associated in a colonial empire.

Some tropical countries have substantial trade barriers, such as tariffs and quotas, that discourage imported forest products from the United States.

Demographics

Some tropical countries have the highest rates of population growth in the world. Even where average per capita income is very low, demand for forest products can be large, particularly in countries where per capita income is growing. In countries where literacy is great and growing, the demand for newsprint and book and writing paper will be great and growing.

Manufacturing Facilities

Many tropical countries have manufacturing facilities for the production of lumber and sometimes plywood, but don't have expensive and sophisticated facilities. Most of the countries in the tropics are net importers of wood products on a value basis, generally because they do not have the resources required to develop pulp and paper facilities and sometimes because of the absence of suitable fiber furnish from indigenous forest tree species.

OPPORTUNITIES FOR REGIONAL TRADE

The opportunities for U.S. forest products industries to sell in the markets of the tropical countries are great—especially for firms willing to make major efforts to understand and develop these markets—because most tropical countries produce wood commodities that are generally not competitive with U.S. wood commodities.

Currently the United States is a major importer of tropical wood commodities; however, if trade between the tropical countries and the United States is to be fostered as a long-term commercial enterprise, it must, in most cases, by mutually advantageous. While some of the oil-rich countries in the tropics may be able to buy U.S. forest products out of oil revenues, many of the poor tropical countries must earn the foreign exchange required to pay for imports by selling other commodities abroad. Often the most available and accessible exportable resource is wood. If the United States wishes to engage in trade with tropical countries, and this should be both a commercially and politically desirable objective, then it may as a nation wish to work with these countries to fashion complementary rather than competitive industries. In the forest products field, given the difference in the nature of the resources, this can often be a feasible goal. It is, however, not something that is likely to emerge spontaneously from the normal operation of the market. It can be encouraged by a carefully crafted combination of technical assistance and a most favored nation trade status. The development of a favorable technological climate for reciprocal trade in forest products is an appropriate role for the federal government.

The United States fosters technical assistance programs in many of the tropical countries that are logical trading partners in forest products. The United States Agency for International Development (USAID) is the federal establishment responsible for administration of the technical assistance program. In recent years, USAID has stated that a major thrust of its program is to strengthen the private sector in the countries that it undertakes to assist. These are predominantly tropical countries. Other countries—including Canada, Japan, Sweden, and Finland—have pursued such programs to their own advantage and that of the recipient countries. This private sector effort has nonetheless been missing from the USAID forestry program. Introduction of such an initiative could make a major contribution to the development of mutually advantageous trade programs in forest products between the United States and a number of tropical countries. The forest products industry and the forestry schools of the United States have much expertise that could be made available to the federal government for such an effort. Among the advantages that can accrue to the U.S. forest products industry, in addition to the structuring of matching industries that can ultimately engage in reciprocal trade, is that such a program helps to provide a cadre of forestry professionals familiar with the countries that are potential trading partners and possessing an understanding of their language, culture, and economic status.

In summary, the tropical countries are both suppliers and consumers of forest products. On the whole they are net consumers and are likely to continue to be. As their populations grow and they continue to develop, their demand for forest products will increase. Many of these countries have the potential to become important customers for U.S. forest products. The federal government could assist this development through well-designed and well-implemented technical assistance programs. It will take patience on the part of the industry and effort on the part of the government to develop these markets, but it will be rewarding in the long run.

Chapter 20

Tropical Countries as Suppliers of Timber to International and Domestic Markets

THEO ERFURTH

TRENDS AND PATTERNS IN TROPICAL TIMBER TRADE

Exports of tropical timber rose from U.S.$3.1 billion in 1973 to $5.3 billion in 1983 (Table 1). Present export levels are estimated to be below $5 billion. World trade in tropical timber reached its peak in 1979-80, when it exceeded the $8 billion mark.

Table 1. Tropical timber exports in world trade of forest products, 1973-83 (U.S.$ million).

Exports	1973	%	1983	%
Forest Products				
Total world	22,227	100	47,254	100
Developing countries	3,854	17	6,805	14
Tropical Timber				
Developing countries	3,133*	100	5,365	100
Logs	1,884	60	2,549	48
Sawnwood	615	20	1,555	29
Veneer	110	4	193	3
Plywood	524	16	1,068	20

*14% of total world exports of forest products.
Sources: FAO *Yearbook of Forest Products*, FAO *Monthly Bulletin of Tropical Forest Products in World Timber Trade*.

The term "tropical timber" refers to logs, sawnwood, veneer, and plywood made of tropical wood species that originated in tropical forests. This meaning corresponds with international usage and is applied throughout this paper.

The tropical timber trade made up 11% of the total world trade in forest products in 1983. The proportion of log exports in total tropical timber exports decreased from 60% in 1973 to 48% in 1983 and has continued to decrease. Sawnwood and plywood exports have been increasing and are expected to continue to do so, particularly plywood from Southeast Asia.

Between 1973 and 1983, plywood exports from Far Eastern developing countries almost doubled, mainly because of the sharply rising plywood production capacities in Indonesia. Often overlooked is the more spectacular increase of the tropical sawnwood trade, which is more evenly spread over the four regions, although Malaysia still accounts for the largest share. As shown in Table 2, tropical sawnwood exports from all four regions have risen from U.S.$615 million in 1973

Table 2. Tropical timber exports by regions and products,* 1973-83 (U.S.$ million).

	Logs		Sawnwood		Veneer		Plywood		Total	
	1973	1983	1973	1983	1973	1983	1973	1983	1973	1983
(a) World total	2,064	2,887	1,123	2,616	443	718	1,403	2,589	5,033	8,810
(b) Four regions**	1,884	2,549	615	1,555	110	193	534	1,068	3,143	5,365
	(91)	(88)	(55)	(59)	(25)	(27)	(38)	(41)	(62)	(61)
Far Eastern developing countries	1,332	1,914	447	1,208	41	77	482	948	2,302	4,147
African developing countries	522	555	100	126	34	66	32	28	688	775
Latin America	17	7	65	214	34	48	17	88	133	357
Oceania	13	73	3	7	1	2	3	4	20	86

*Nonconiferous logs and sawnwood, veneer, and plywood, all species.
**Figures in parentheses are percentages of (b) in (a).
Sources: FAO Yearbook of Forest Products, FAO Monthly Bulletin of Tropical Forest Products in World Timber Trade.

to $1,555 million in 1983. In 1983 tropical sawnwood accounted for 59% of the total world exports in broadleaf (nonconiferous) sawnwood. On the other hand, the total value of tropical log exports was still comparatively high: the combined export values of the three other tropical timber products—sawnwood, veneer, and plywood—were at the same level as log export values alone. However, in recent years log exports have declined notably. The percentage of tropical timber exports, as shown in the right-hand columns of Table 2, in the world total was virtually the same in 1973 and 1983—62% and 61%—despite the notable shifts within the categories (logs, sawnwood, veneer, and plywood).

SAW AND VENEER LOG PRODUCTION: WOOD SPECIES AND SUPPLY PROSPECTS

For foresters working in temperate zone forests, knowledge of and experience with individual tree species are normal prerequisites for doing the job. A forester dealing with tropical forests may find that such detailed knowledge by species becomes an unwieldy and highly demanding, rather scientific addition to his basic education, which he sometimes acquires in different environments. The intricate character of all aspects of tropical forestry, because of the complexity of its botanical composition, aggravated by special problems such as the vulnerability of tropical soils, difficult accessibility, and the effects of rainy seasons, explains why tropical forest management, of which wood harvesting is an essential part, is so demanding of those involved.

The data presented in Table 3 are part of a comprehensive analysis of more than a thousand wood species. The table is largely self-explanatory, so a few additional remarks will suffice. Species grouping in Southeast Asia is fairly standard within families and genera, but the number of species under one pilot name may vary from country to country and from case to case. The ten pilot or trade names under which the main commercial species for each region are listed account for more than half of the total log production: in the case of West and Central Africa, 59%; Southeast Asia, 73%; and tropical South America, 50%.

In accordance with the criteria established for the purpose of the study, the estimated numbers of commercial and lesser used species, by region, are as follows: West and Central Africa, 105 commercial and 112 lesser used species; tropical South America, 210 commercial and 263 lesser used species; Southeast Asia—in view of the special situation in this region—290 commercial dipterocarp species, 88 lesser used dipterocarp species, and in the nondipterocarp category, 311 commercial and 377 lesser used wood species.

In the course of time, various terms have been employed to characterize insufficiently used wood species, such as secondary species, lesser known wood species, weed species, and little used species. None of these terms explains the circumstances that make a particular species lesser known, little used, or secondary. There are several reasons it is difficult to arrive at generally valid definitions; for instance, it may happen that a particular species is little used in one country and fully commercialized in another, or that insufficient statistical coverage hampers comparability. An essential prerequisite for promotional activity is information on not only the qualitative but the quantitative or supply aspects of individual or groups of species. Basic knowledge on wood species should be sufficiently comprehensive to allow conclusions as to whether promotional activity is worthwhile.

A paper I presented to the conference on International Forest Products Trade: Resources and Market Opportunities, organized by the Forest Products Research Society, held in 1983 in Arlington, reviewed in some detail the supply prospects for tropical timber (Erfurth 1983). The conclusions are still valid, since revised basic data are not available. I extract some relevant parts:

1. Estimates and projections to the year 2000 roughly indicate a doubling of net removals of saw and veneer logs in Africa, Asia, and Oceania, and a threefold increase in Latin America.

2. Because of the projected relatively fast expansion of local processing and domestic consump-

tion, the proportion of utility timbers, lesser used species, and lower quality wood in the overall wood use and consumption pattern is expected to increase. The demand for fine and specialty timber will in some cases outstrip supplies, possibly resulting in higher prices for species in great demand, or in the substitution of certain species by other wood or nonwood materials.

3. There will be a further decrease of natural forest areas with all the implications this entails in the longer term. However, supplies of tropical logs up to the year 2000 are expected to meet requirements. This situation will also depend on adequate action being taken to increase utilization of lesser used species and lower quality logs, to expand processing facilities and development of infrastructure, particularly within and to more remote forests, and to carry out product research and development, market and product promotion, and related institutional arrangements.

4. Outlook studies that include tropical timber are not yet available for estimates of supplies beyond the year 2000. Generally, it can be assumed that future world consumption of wood and wood products will exceed the level reached by the year 2000 and also that the demand for tropical timber will further increase.

5. If the area of natural tropical forests continues to decrease or degrade at a rate similar to that experienced until the year 2000, the supplies of tropical timber, especially for international markets, will be seriously affected. More effective measures need to be taken as early as possible to improve forest management and to expand reforestation. Otherwise both exports and domestic consumption—which are increasingly complementing each other—will not play the role they could play, and are expected to play, in economic and social development.

WOOD PROPERTIES, COMMERCIAL ACCEPTANCE, AND PRODUCT DEVELOPMENT

Two factors account for the variability of tropical timber properties: the heterogeneous species composition of the tropical forests, and the property variations and defects within individual wood species. In no other forest type are heterogeneity and variability so pronounced. Moreover, the complex botanical composition and numerous wood species of various ages means that wood supplies available in each forest area are highly diversified in properties, forms, and sizes.

On the other hand, promoters of wood often emphasize its versatility and its competitiveness in relation to other materials. How far is this true for tropical timber? To answer that question it is important to avoid generalizations. Before judgments are made, one should look into costs, prices, use properties, and intrinsic wood values, and their realization in specific uses. Timber of high value or use properties can fetch a premium price when actual production and transport costs account for only a fraction of that price. When market prices are far below production and transport costs, producers will not hesitate to eliminate the timber, which then enters the category of lesser used species, some of which may be lesser known to the operator and some may just be of low quality.

There are basically three approaches for investigating fuller utilization: (1) the possibilities of optimizing wood values of individual woods, (2) the grouping of those wood species that do not qualify for consideration under item 1, and (3) the promotion of "mixed tropical hardwoods" for integrated industrial use in the form of chips and fibers.

The success or failure of grouping wood species for promotional purposes depends largely on the degree of similarity in use properties. The closer use properties of different species are, the greater the chances for their market acceptance, perhaps even for more sophisticated uses. However, a few deviations from average properties—for example, in color or surface finishing—may cause severe disturbances in the chain linking producers, traders, and users. Larger differences seriously and adversely affect prices and result in breaking up the grouping; and the consequence is that those species with deviating or varying properties have to be singled out. To some extent utility grades and similar assortments might tolerate deviations from a given average quality, but at

Table 3. Saw and veneer log production of principal commercial tropical timbers by major regions and producing countries, ca. 1973 (1,000 m³).

WEST AND CENTRAL AFRICA

Production by Country**

Pilot Name	Scientific Name	No. of Possible Species*	Log Production	LIB	IVC	GHA	NIG	CAM	GAB	CGO	ZAI
1. Obeche	Triplochiton scleroxylon	1	1,960(16)†	21	1,109	520	229	69	–	5	7
2. Okoumé	Aucoumea klaineana	1	1,782(15)	–	–	–	–	–	1,571	211	–
3. Sipo	Entandrophragma utile	1	661(5)	49	430	107	2	26	9	7	31
4. Acajou-mahogany	Khaya spp.	3	586(5)	3	195	161	136	18	19	28	26
5. Sapelli	Entandrophragma cyclindricum	1	552(4)	7	172	190	7	84	–	61	31
6. Limba	Terminalia superba	1	462(4)	–	51	1	265	16	39	65	24
7. Iroko	Chlorophora excelsa	1	349(3)	9	184	95	17	18	–	6	20
8. Tiama	Entandrophragma angolense	1	308(3)	2	179	83	6	4	5	3	26
9. Azobe	Lophira alata	1	284(2)	16	5	1	1	261	–	–	–
10. Makore-douka	Tieghemella spp.	2	274(2)(59)	20	166	56	–	2	15	15	–

SOUTHEAST ASIA

Pilot Name	Scientific Name	No. of Possible Species*	Log Production	THAI	MAL	SAR	SAB	IDO 1973	PHIL 1971
1. Meranti‡	Shorea 92, Parashorea 3	95	17,688(27)	–	3,312	455	–	13,901	–
2. Philippine mahogany	Shorea 5, Parashorea 1	7	7,084(11)	–	–	–	–	–	7,084
3. Red seraya	Shorea spp. Pentacme 1	32	5,934(9)	–	–	–	5,934	–	–
4. Keruing	Dipterocarpus spp.	52	4,749(7)	–	2,042	85	1,205	1,417	–
5. Kapur	Dryobalanops spp.	9	3,516(5)	–	1,313	85	1,867	251	–
6. White seraya	Parashorea spp.	5	3,091(5)	–	–	–	3,091	–	–
7. Ramin	Gonystylus spp.	8	2,748(4)	–	–	640	1	2,107	–
8. Apitong	Dipterocarpus spp.	10	1,597(2)	–	–	–	–	–	1,597
9. Alan	Shorea albida	1	1,399(2)	–	–	1,399	–	–	–
10. Yellow seraya	Shorea spp.	10	949(1)(73)	–	–	–	949	–	–

TROPICAL SOUTH AMERICA

Pilot Name	Scientific Name	No. of Possible Species*	Log Production	BRA 1972	BOL 1973	PER 1973	ECU 1973	COL 1971	VEN 1972	GUY 1973	SUR 1973	GUF 1971
1. Virola	Virola spp.	7	1,662(24)	1,105	3	0.8	5	510	–	3	35	0.4
2. Caoba	Swietenia macrophylla	1	411(6)	204	150	34	–	–	23	–	–	–
3. Cedro	Cedrela spp.	2	265(4)	107	0.7	130	13	12	–	–	0.3	2
4. Balsa	Ochroma lagopus	1	246(3)	–	–	–	246	–	–	–	–	–

			Total (%)†							
5.	Andiroba	Carapa spp.	2	219 (3)	151	--	13	34	--	5	16	0.4
6.	Sajo	Campnosperma panamensis	1	182 (3)	--	--	--	182	--	--	--	--
7.	Louro inhamuy	Ocotea cymbarum	1	149 (2)	149	--	--	1	124	--	--	--
8.	Saqui-saqui	Bombacopsis quinatum	1	125 (2)	--	--	--	--	--	--	--	--
9.	Greenheart	Ocotea rodiaea	1	117 (2)	--	--	--	--	--	117	--	--
10.	Mijao	Anacardium excelsum	1	106 (1)(50)	--	--	6	8	92	--	--	--

* Species grouping in Southeast Asia may involve double employment of individual wood species.

** All production data are for the year 1973, except for tropical South America.

† Percentages of regional totals in parentheses.

‡ Includes dark red, light red, white, and yellow meranti from Malaysia (Peninsula and Sarawak), and *Meranti merah* (red) and *Meranti putih* (white) from Indonesia.

ABBREVIATIONS:

West and Central Africa	Southeast Asia	Tropical South America
LIB - Liberia	THAI - Thailand	BRA - Brazilian Amazon
IVC - Ivory Coast	MAL - West Malaysia	BOL - Bolivia
GHA - Ghana	SAR - Sarawak	PER - Peru
NIG - Nigeria	SAB - Sabah	ECU - Ecuador
CAM - Cameroon	IDO - Indonesia	COL - Colombia
GAB - Gabon	PHIL - Philippines	VEN - Venezuela
CGO - Congo		GUY - Guyana
ZAI - Zaire		SUR - Suriname
		GUF - French Guiana

Source: FAO (1976).

the expense of price. Grouping of wood species has so far been practiced on a commercial scale only in Southeast Asia with traditional trade names such as meranti, seraya, lauan, and mainly those that belong to the dipterocarp family.

Integrated industrial use has its merits, in that certain tropical wood species will in future not, or only insufficiently, be used in their solid form. Most of these species can be, and are in some countries, used in sizable quantities in their disintegrated form as chips and fibers for board, pulp, and paper manufacture. In some cases, large portions of locally produced chips are being exported for processing overseas. From the forester's point of view, such massive production of chips presents a rather radical solution to the lesser used species problem, since the practice has frequently led to aggressive forest exploitation and denudation. Conscientious land use planning and management need to be the basis for these kinds of activities.

Let us go back to item 1, the possibilities of optimizing wood values. Added value through product development is the backbone of progress in this field. Careful attention needs to be given to the level of wood value at which new value is added through further processing. In this scale of high to low wood values, the actual processing cost may represent only a small fraction of the price for, say, valuable decorative veneer species; or, as in the case of building timber, the processing cost may be a large share of the product price, and thus the relative share of added value through local processing may be high in the total cost. This line of thought needs to be followed in all cases of product development in order to determine, in the light of market investigations, how and where the available timbers should be processed.

SUPPLIES TO INTERNATIONAL MARKETS

Tropical timber supplies to traditional overseas markets are usually classified as follows: (1) *Fine timbers* roughly comprise those species that combine decorative values with medium to good strength and working properties. Color, grain, and form stability are usually decisive factors for uses such as cabinetmaking and high quality furniture and joinery, decorative panels, and veneers. Examples are mahogany, mahogany type timbers, and teak. (2) *Specialty timbers* usually combine good strength properties with one or several other outstanding characteristics such as durability in contact with the ground or underwater, bursting strength, and so forth, for uses such as marine piling and construction, stair treads, shingles, outside joinery, and boatbuilding. Examples are greenheart, iroko, bongossi, merbun, apitong, and keruing. (3) *Utility timbers* have properties that allow a wide range of uses and applications. Tropical species usually have large diameters, are relatively free of defects (knots), and have other characteristics that make them competitive with, or complementary to, utility timbers growing in the consuming countries. Uses are in the fields of general construction, utility veneer, and plywood, including corestock, packaging, low quality furniture, and so on.

This brief analytical division reflects the *qualitative* requirements of importing countries. The *quantitative* aspects of tropical timber imports for some selected countries are given in Table 4.

This table brings out, in roundwood equivalents, tropical timber imports in relation to the respective total industrial wood consumption. Korea and Japan, with their large dependency on log imports, are in contrast to the United States and Australia, which import processed tropical timber almost exclusively. France and Italy represent import patterns typical for West European countries with the exception of the United Kingdom, where log imports are of relatively small importance, and Greece, Portugal, and Spain, where utility qualities are predominant. Utility type timbers for the mass manufacture of plywood also prevail in the imports of Japan and Korea. A large portion of the tropical timber imported by the United States is used for thin "hardwood plywood," or veneer sheets, often for decorative purposes. U.S. sawnwood imports contain a sizable proportion of fine and specialty timbers, and the United States is the largest importer of further-processed tropical timber products.

Table 4. Tropical imports of selected major consuming countries in relation to industrial wood consumption, 1980 (1,000 m^3).

	United States	Japan	France	Italy	United Kingdom	Australia	South Korea
CONSUMPTION							
Coniferous sawnwood	75,822	35,596	8,091	5,475	7,032	1,320	1,212
Nonconiferous sawnwood	16,830	6,894	4,517	2,793	1,265	2,204	1,389
Veneer and plywood	16,827	8,733	865	947	807	194	668
Total in roundwood equivalents*	195,657	91,951	23,660	16,199	15,865	7,479	6,075
IMPORTS OF TROPICAL TIMBER							
Logs	11	18,989	1,640	1,290	96	--	4,582
Sawnwood	416	513	615	493	404	276	23
Veneer and plywood	1,012	77	210	37	325	54	(40)
Total in roundwood equivalents*	3,054	20,091**	3,225	2,268	1,570	622	4,704
Imports of tropical timber as percentage of consumption	1.6	21.8	13.6	14.0	9.9	8.3	77.4
(Excluding coniferous sawnwood)	(4.4)	(61.8)	(31.8)	(32.1)	(38.1)	(14.0)	(116.1)†

*Conversion factors used to calculate roundwood equivalent (m^3 of roundwood for 1 m^3 of end product): coniferous sawnwood, 1.67; nonconiferous sawnwood, 1.82; veneer 1.90; plywood, 2.30.
**Excluding 132,000 m^3 pulpwood (in round) and 461,000 tons of chips and particles imported from tropical countries.
†Above 100% because of reexport.
Source: FAO *Yearbook of Forest Products*.

In West European countries the use patterns are probably more diversified than elsewhere, but uses in the fine and specialty timber categories prevail. On the whole, the role of tropical timber imports into the United States and Western Europe is one of complementing what is available in the countries themselves. Thus market acceptance is a function of the competition between locally available and imported wood products.

There is no doubt that aggressive promotion and new efforts in product development and improvement—offered in collaboration with importing countries—have led to increasing international trade of more processed tropical timber products, though only a few countries have participated in these efforts. To the producer-exporter, this higher degree of processing means that he is entering new technical and economic grounds, where he will have to employ new methods in order to achieve his new objectives.

For instance, changing from one product to another, more processed one inevitably implies different and more exacting production methods and marketing practices. The problem really is how the semiprocessed product can be more readily integrated into the respective stages of the customers' production lines, or the finished product into final application. Since tropical forest products usually have to be transported over long distances, sometimes taking several months, specific problems arise in transporting the goods from producer to consumer. In this context the timing of delivery to the overseas customer needs special attention, but generally the problems that receive attention are related to product specification and standardization, dimensional accuracy, tolerances and quality control, and, last but not least, the type and quality of packaging.

It is not the intention in this paper to go into details of *industrial* development subjects and problems, nor to deal with institutional aspects in general. However, industries and trade, as the operational and organizational entities, need to be supported in their development efforts as early as possible. Early involvement and mutual support of both the public and the private sectors, particularly in the producing countries, are essential for progress in socioeconomic development. This last observation leads to the discussion of the development of domestic markets.

THE DEVELOPMENT OF DOMESTIC MARKETS

The term "developing country" is frequently used today to indicate that further progress is needed for the country to reach a satisfactory standard of living and related industrialization. "Threshold countries" are those that are close to this objective. However, the majority of developing countries are still below this threshold at various levels of socioeconomic development.

There are also many different levels of wood use—from both the quantitative and qualitative standpoints. These differences are the result of a variety of factors, which need to be identified for each case. More important factors are the kind and size of forest resources, the traditional attitudes toward wood use and woodworking, the density and distribution of the population as well as their socioeconomic position, and the transportation infrastructure.

There is no single recipe for developing domestic markets. Rural and urban areas, private and public sector requirements, need to be analyzed in connection with the above factors. This analysis, in turn, can provide the starting point for action in a specific locality.

Before going any further, let me briefly analyze recent trends in domestic processing. As shown in Table 5, considerable progress has been made between 1973 and 1983. This is particularly true of Southeast Asia, the largest producer of tropical timber; but progress has not been shared equally by all countries and varies also within regions.

Tropical Latin American countries have traditionally converted most of their forest removals locally, and several of them have prohibited log exports. It is no surprise that by 1983 virtually 100% were converted locally in the countries concerned (as shown in Table 5). Southeast Asian countries by 1983 had processed locally more than two-thirds of their saw and veneer log production, and tropical African countries reached almost the same proportion. Generally speaking, domestic con-

Table 5. Tropical saw and veneer log production and domestic processing, 1973-83 (million m^3).

Region	Saw and Veneer Log Production		Domestic Processing in Roundwood Equivalents		Percentage* of Domestic Processing	
	1973	1983	1973	1983	1973	1983
Far Eastern developing countries	78.0	82.0 (est.)	40.0	56.0	51.0	68.0
African developing countries	17.0	18.0	7.0	11.0	41.0	61.0
Latin America	19.0	28.0	18.0	28.0	94.0	100.0**
Oceania	1.2	1.6	0.5	0.5	40.0	27.0

*Differences due to rounding.
**Tropical log exports in 1983 were 46,000 m^3.
Source: FAO *Yearbook of Forest Products*.

sumption has also increased. There is a conviction in tropical forest countries that timber can and should play a more important and direct role in socioeconomic development.

Unlike export markets, where product types and qualities are usually limited, in domestic markets the demand—present and potential—comprises the full range of forest products and their end uses. Appropriate product choice and reliability in product performance at acceptable costs and prices are the key for opening and expanding new market sectors. More emphasis needs to be put on product planning in relation to market requirements and wood availability before the establishment or adjustment of product lines. Unless wood products are in a position to meet the (possibly changing) requirements in quality, condition, and form, they will be unable to compete with other, possibly imported materials.

Urban centers are expanding fast in many developing countries, and this creates demands for building materials, furniture, and numerous other articles for which wood competes with other materials. Efforts to increase the flow of timber products to urban centers need to take into account, at an early stage, the market-oriented socioeconomic, technical, and organizational aspects. Local rural enterprises usually do not or cannot provide technical services and essential know-how to consumer areas. Cooperative efforts between consumers and producers, supported by development agencies, can in these cases complement the more technical aspects of promotion and marketing activities.

Considerations of economic geographic and the related cost structure often play an important role in decision making in marketing. Product distribution, transportation infrastructure, and the location of intermediate storage facilities and processing plants all need to be considered along with the other issues mentioned above. Examples of such interrelated operations are of wide interest in other developing areas. The scarcity of such cases or examples suggests the need to collect and evaluate more relevant data for the benefit of those directly interested in development.

Any effort to promote wood use in rural development would have to build on existing levels of wood use and the available local skills to work and handle wood. Traditional approaches to wood use in housing, for instance, may vary considerably even if similar kinds of wood are available. The lever for wood promotion should be at the point where direct improvements in living conditions can be expected. One example: the use of improved or new types of roof construction in rural housing as a first step in introducing other housing features such as doors, windows, and wall elements, along with support or extension services and training at the "grass roots level." In other areas where wood use is already more common, the introduction of suitable wood seasoning and preservation methods might suffice to stimulate the wider and better use of wood. It is obvious that in each case the selection of the starting point for promotion is important to get the chain of action moving.

INTERNATIONAL SUPPORT ACTIVITIES

International cooperation and assistance in the field of tropical forests and wood utilization have been given early attention by governments and international organizations. Shortly after its foundation, the Food and Agriculture Organization of the United Nations (FAO) published in its first issue of *Unasylva* (1947) an article by A. Aubréville, of France, on "The Disappearance of the Tropical Forests of Africa," which sought adequate measures to protect soil and forest cover in developing African forests.

The second article in this first issue of *Unasylva* was written by T. Alfred Hall, then director of the Pacific Northwest Forest and Range Experiment Station of the U.S. Forest Service. He concluded his article on "Forest Utilization," which included tropical forests and timber in a worldwide review, as follows: "There has been inadequate opportunity for the transferring of scientific knowledge . . . into actual operating industries with full protection of public interest and opportunity for private gain. FAO should be in an excellent situation to bridge this gap, to adduce all available information on the subject and put it to work in the major task of drawing goods from the world's renewable forest wealth."

Forty years have passed since then. FAO's forestry activities, which indeed started as a kind of world center for information exchange, have expanded into what is today the largest international development agency in forestry. New problems have emerged, but many of the early preoccupations concerning the world's forests have remained, some have disappeared, and some have grown larger, requiring new initiatives and solutions.

In the meantime, other organizations have been created which closely collaborate with FAO, such as UNEP (United Nations Environmental Program), UNIDO (United Nations Industrial Development Organization), UNCTAD (United Nations Conference on Trade and Development), ITC (International Trade Center; UNCTAD/GATT), and GATT (General Agreement on Tariffs and Trade). Directly concerned with financing development projects are the World Bank (IBRD) and the United Nations Development Programme (UNDP). There are many other governmental and nongovernmental organizations at the world and regional level with specific development tasks in tropical forestry. It is not possible to cover them all in this paper, but two new activities need to be mentioned here, since they are directly concerned with tropical forests and tropical timber.

The Tropical Forestry Action Plan 1985 is the result of concerted efforts that developed in the FAO Committee on Forest Development in the Tropics. Its objectives are, in brief, (1) strengthening coordination between the organizations and countries concerned, (2) developing further and supporting a development strategy for action, and (3) identifying investment needs. The World Bank, the United Nations Development Programme, and the World Resources Institute collaborated in its preparation.

The Plan identifies five priority areas: (1) forestry in land use, (2) forest-based industrial development, (3) fuelwood and energy, (4) conservation of tropical forest ecosystems, and (5) institutions. Action in the second area aims at promoting appropriate forest industries in an integrated way through intensification of resource management and development, appropriate raw material harvesting, establishment and management of appropriate forest industries, reduction of waste, and development of capability in marketing forest products.

The International Tropical Timber Organization (ITTO) became operational in 1986. The ITTO council at the conclusion of its first session appointed an executive secretary and decided that Yokohama, Japan, would be the seat of the organization. Negotiations had started in 1977 between producing and consuming countries of tropical timber within the framework of the UNCTAD Integrated Program for Commodities and in close collaboration with FAO. After a series of preparatory meetings, the text of the Agreement was adopted by United Nations Conference on Tropical Timber in Geneva, in which representatives of sixty-five countries participated.

The parties to this agreement recognized the importance of conservation and development of tropical forests and of tropical timber to the economies of member countries. The Agreement, in brief, includes the following objectives: (1) provision of an effective framework for cooperation and consultation between tropical timber producing and consuming members, (2) expansion and diversification of international trade in tropical timber and improvement of structural conditions in the tropical timber market, (3) support of research and development with a view to improving forest management and wood utilization, (4) improvement of market intelligence with a view to ensuring greater clarity in the international tropical timber market, (5) encouragement of further processing of tropical timber in producing member countries with a view to promoting their industrialization and thereby increasing their export earnings, (6) encouragement of members to support and develop industrial tropical timber reforestation and forest management activities, (7) improved marketing and distribution of tropical timber exports of producing members, and (8) encouragement of the development of national policies aimed at sustainable utilization and conservation of tropical forests and their genetic resources, and maintaining the ecological balance in the regions concerned.

REFERENCES

Aubréville, A. 1947. The disappearance of the tropical forests of Africa. Unasylva 1(1):5-11.

Erfurth, T. 1983. Trends in timber supplies from tropical regions. *In* International forest products trade: Resources and market opportunities. FPRS proceedings, Madison, Wisconsin.

FAO. 1976. The marketing of tropical wood. FO:misc/76/8. Food and Agriculture Organization, Rome.

——. 1986. Yearbook of forest products. Food and Agriculture Organization, Rome.

Hall, J. A. 1947. Forest utilization. Unasylva 1(1): 12-26.

Chapter 21

Tropical Hardwood Supplies in the South Pacific

F. A. FLYNN

The South Pacific timber producing countries are Australia, Fiji, New Caledonia, New Zealand, Papua New Guinea (PNG), Samoa, Solomon Islands, and Vanuatu. For the purposes of this paper Australia and New Zealand are excluded because their tropical timber production has been reduced to insignificant levels. Since most of their tropical forest areas have been converted into national parks, they have no potential as exporters of tropical hardwoods. New Caledonia is excluded because it has no record of exporting tropical hardwoods and it imports more timber than it produces.

The remaining South Pacific countries have a closed forest area of 38,134,000 hectares (ha) compared with the combined area of Indonesia, Malaysia, and the Philippines of 157,001,000 ha. The South Pacific countries will be referred to in this paper as the Islands, and Indonesia, Malaysia, and Philippines as ASEAN (as members of the Association of Southeast Asian Nations). Thus the Islands have 19.5% of the combined closed forest area. However, the Islands produce only 3.5% of the combined roundwood production.

The question is: for the purposes of future supply have the ASEAN resources been overexploited or the Islands underutilized? It appears that the answer is somewhere in between. The Islands have relatively low population densities compared with ASEAN, and there has not been the same pressure to clear land. It was only in the late 1970s that Papua New Guinea and the Solomon Islands developed their log export trade to any extent.

CAPACITY FOR EXPORT

Since each of the Islands has completely different characteristics, they will be discussed separately in alphabetical order. Statistical information concerning consumption, production, importation, and exportation of selected wood products from the Islands for the period 1973 to 1984 is given in an appendix.

Fiji

Of the island countries only Fiji has had a national forest resource survey, which was carried out in the late 1960s. This information has been used to implement a forest policy the basis of which is a vigorous reforestation program, incentives to the conversion industry to modernize equipment, encouragement to export processed products, and strict control of log exports. In many ways Fiji is a model of what can be done in tropical island countries.

Reforestation in Fiji has been of three different types: (1) exotic softwoods, (2) exotic hardwoods, and (3) indigenous hardwoods.

Exotic Softwoods. Fiji has planted grasslands with over 50,000 ha of *Pinus caribaea* and has plans to provide, on a perpetual basis, over 1 million m^3 of logs per year from this resource. A joint venture between British Petroleum and the Fiji government is currently building a $50 million saw-mill, chip mill, and power generation plant based on this resource.

Although the purpose of this paper is to discuss tropical hardwood supplies, the above information has a direct bearing on this subject, since the pine will be used as a building grade timber, thus freeing indigenous production for export.

Exotic Hardwoods. Fiji has been reforesting with exotic hardwoods for the last twenty-five years, primarily mahogany (*Swietenia macrophylla*). Although there was some concern that quality would be affected by ambrosia attack, recent extensive peeling and sawing trials have revealed this is minimal, and as a consequence planting has been accelerated to 5,000 ha/yr. Rotation appears to be around 50 years, and as such mahogany will be a profitable crop. Sawn production of exotic hardwoods is expected to reach 80,000 m^3 by the year 2000. In addition, Fiji's plywood operation is expected to manufacture high quality marine grade plywood for export.

Indigenous Hardwoods. Replanting of indigenous species has also been taking place, of which the main species has been kauvula (*Endospermum macrophyllum*). Trials have also been carried out with dakua makadre (*Agathis vitiensis*), otherwise known as kauri pine. Logging of the indigenous forest is strictly controlled, and all trees less than 35 cm diameter at breast height are left behind for a further crop.

Fiji's capacity to export tropical hardwoods will increase rapidly between now and the year 2000, with availability of indigenous timbers increasing from 18,000 m^3/yr to 44,000 m^3/yr and exotic hardwoods (mainly mahogany) increasing from 2,000 m^3/yr to 80,000 m^3/yr over the same period. The strategy will be to add as much value to the timber as possible. Sufficient kiln drying capacity is being added to dry Fiji's total exports. Modern machining facilities are also being added for the production of moldings and paneling.

Production of high quality plywood, particularly marine grade, will be expanded. The intention is to find niches not filled by the large Indonesian producers. Exports are expected to be no more than 5,000 m^3/yr.

Fiji has already earned a reputation in the United States, Australia, New Zealand, and Japan for producing high quality thin rotary face veneer. This is also niche marketed into higher priced areas. Exports of rotary veneer are expected to grow at a modest rate from current levels of 5,000 m^3/yr. Growth in veneer exports is likely to come from sliced veneers, with two new plants expected to be in operation within the next five years.

Papua New Guinea (PNG)

Of the island countries PNG is obviously the sleeping giant as far as potential is concerned. Its forest area exceeds that of Malaysia and the Philippines combined; however, sawnwood production was only 124,000 m^3 in 1984, compared with 8,363,000 m^3 for those two countries, or less than 2%. PNG is estimated to have 250 years' supply of indigenous logs at a consumption rate of 2 million m^3/yr. Clearly, with sound forest policies, PNG can be developed as a major supplier of tropical hardwoods.

Unfortunately, all development seems to be in log exports, which have trebled from 445,000 m^3 in 1979 to 1,300,000 m^3 in 1984. During the same time, sawn production declined from 138,000 m^3 to 124,000 m^3. Export sales of sawn timber dropped from 47,000 m^3 to 22,000 m^3 for the same period, despite government incentives to export processed timber rather than logs.

FAO consultant Karl H. Kehr, in his review of the industry (1986), states that "new sawmilling capacities are badly needed as the local market should be supplied reliably in the future and sawn timber prices be controlled by free market dynamics and a supply/demand relationship."

The PNG government has formed a State Marketing Agency with the right of first refusal for 25% of logs produced for export. This, together with incentives to export processed products instead of logs, may arrest the trend of steeply rising log exports and falling sawn timber production and export.

Plantations totaling 21,700 ha were established in PNG by 1980, of which 3,700 ha were slow growing hardwoods (mainly *Tectona grandis*), 4,800 ha of fast growing hardwoods (mainly *Eucalyptus robusta* and *E. deglupta*), and 13,200 ha of softwoods (*Araucaria hunsteinii*, and *A. cunninghamii*, *Pinus patula*, and *P. caribaea*).

PNG's plywood industry is in decline, with production falling from 15,000 m^3 in 1979 to 9,000 m^3 in 1984 while exports remained steady for the same perod at 6,000 m^3/yr.

Obviously Papua New Guinea has tremendous potential for increasing exports of tropical hardwoods and is attempting to lessen its reliance on log exports and increase production of processed products.

Samoa

Samoa has the smallest resource base of the island countries, with an area of closed forest of 142,000 ha. It is the only one of the countries to import relatively substantial quantities of its sawn timber requirements. In 1984 it produced 21,000 m^3, imported 17,000 m^3, and exported 3,000 m^3. In recent years it has installed a rotary veneer plant, and small exports have gone made mainly to Australia. Samoa's resource is too small for it to be considered a major future supplier of tropical hardwoods.

Solomon Islands

The Solomon Islands have the second largest forest area of the Island countries, with a closed forest of 2,454,000 ha. They have the problem of a small home market that has restricted sawmill development, resulting in approximately 94% of log production being exported. Since log exporting is one of the major export earners, there is a real problem as to how to restrict log production to a sustainable level and yet maximize earnings of much needed foreign exchange. It is extremely difficult to build a manufacturing base when the home market is so small. In the case of the Solomons, domestic consumption of sawn timber is around 10,000 m^3/yr.

The government plans for indigenous log extraction to continue at about 400,000 m^3/yr until 1995 and then to decrease to 300,000 m^3/yr for the last five years of this century. Logs converted into sawn timber are planned to increase steadily from 258,000 m^3 for the five years to 1995 to 850,000 m^3 for the five years to the year 2000. (Projection figures provided by the Solomon Islands Forestry Department.)

The Solomon Islands had established 2,385 ha of indigenous forest by the end of 1985, with plans to increase this to 28,385 ha by the year 2000. Species being planted are *Gmelina* and mahogany.

The Solomon Islands will remain a reasonably significant exporter of tropical hardwood logs until the year 1995, when processed products are expected to become increasingly available.

Vanuatu

Vanuatu has a closed forest area of 241,000 ha, making it the second smallest of the group under review. In the five years to 1985 it extracted 64,000 m^3 of indigenous logs and exported 50%. In the the five years to 1990 Vanuatu plans to increase indigenous log extraction to 166,000 m^3 and log exports to 80,000 m^3. Log exports are restricted to one species, milktree (*Antiaris toxicaria*). A ban on log exports is planned for 1990.

Vanuatu has established 1,532 ha of exotic forest and plans to increase this to 8,757 ha by the year 2000. Species planted are *Cordia alliodora* and *Pinus caribaea*. The government plans to export cordia logs to Asia after the year 2000. There are two veneer slicing plants under construction, with a planned production rate of 8,000 m^3 (logs converted), most of which will be for export. (Projection figures provided by the Vanuatu Forestry Department.)

Thus Vanuatu will be a small exporter of logs, sawn timber, and veneer in the foreseeable future.

SUMMARY

The Islands have been relatively small exporters of tropical hardwoods compared with ASEAN. But with good forest planning, such as has been achieved in Fiji, they have the potential to be long-term suppliers.

REFERENCES

Alston, A. S. 1982. Timbers of Fiji: Properties and potential uses. Department of Forestry, Fiji.

FAO. 1984. Yearbook of forest products, 1973-84. Food and Agriculture Organization, Rome.

Kehr, K. H. 1986. Papua New Guinea: A review of the sawmilling industry. FAO consultant.

Ministry of Forests, Fiji. 1985. Timber profile: A review of the forest and timber industries and a programme for development.

Yabaki, K. T. 1986. Uses and conservation of forest resources in the South Pacific. Conservator of Forests, Fiji.

APPENDIX 1. Tropical hardwood statistics of the South Pacific Islands compared with those of ASEAN (Indonesia, Malaysia, and the Philippines).

Sawlogs and Veneer Logs	1973	1974	1975	1976	1977	1978	1979	1980	1981	1982	1983	1984
Production: South Pacific Islands (1,000 m³)												
Fiji	137	159	149	155	155	155	192	234	208	166	181	181
Papua New Guinea	804	889	939	1,181	965	1,186	1,186	1,164	1,080	1,159	1,159	1,159
Samoa	62	62	58	23	43	50	58	58	58	58	58	58
Solomon Islands	265	226	228	264	286	249	302	302	302	302	302	302
Vanuatu	12	12	12	5	6	6	6	6	11	9	14	14
Total	1,280	1,348	1,386	1,628	1,455	1,646	1,744	1,764	1,659	1,694	1,714	1,714
Percentage to ASEAN	2.1	2.6	3.2	2.8	2.5	2.6	2.9	2.8	3.0	2.9	2.7	2.8
Production: ASEAN (1,000 m³)												
Indonesia	26,297	23,210	16,200	23,800	22,900	27,300	25,500	28,109	23,664	22,773	25,833	26,370
Malaysia	24,055	21,498	18,979	26,596	27,588	28,514	28,526	28,526	26,569	30,327	33,010	31,600
Philippines	10,423	7,332	8,441	8,712	7,927	7,169	6,578	6,351	5,400	4,514	4,430	3,849
Total	60,775	52,040	43,620	59,108	58,415	62,983	60,604	62,986	55,633	57,614	63,273	61,819
Exports: South Pacific Islands (1,000 m³)												
Fiji	0	0	0	0	0	0	0	0	4	2	2	2
Papua New Guinea	425	655	372	445	402	445	445	642	749	1,063	1,019	1,300
Samoa	0	0	0	0	0	0	0	0	0	0	0	0
Solomon Islands	254	211	208	241	265	221	258	226	296	277	288	284
Vanuatu	12	1	1	1	1	5	5	5	5	5	4	4
Total	691	867	581	687	668	671	708	873	1,054	1,347	1,313	1,590
Percentage to ASEAN	1.8	2.6	2.1	1.9	1.8	1.7	2.0	2.8	4.4	5.6	5.7	8.1
Exports: ASEAN (1,000 m³)												
Indonesia	18,500	16,873	12,884	18,105	18,932	19,457	18,161	15,182	6,489	3,220	3,091	1,646
Malaysia	12,887	12,176	10,793	15,505	16,118	16,717	16,500	15,151	15,870	19,301	18,806	16,676
Philippines	7,759	4,693	4,596	2,331	2,047	2,200	1,248	1,154	1,683	1,590	1,017	1,323
Total	39,146	33,742	28,273	35,941	37,097	38,374	35,909	31,487	24,042	24,111	22,914	19,645

APPENDIX 1. Continued.

Sawnwood and Sleepers	1973	1974	1975	1976	1977	1978	1979	1980	1981	1982	1983	1984
Production: South Pacific Islands (1,000 m³)												
Fiji	82	95	72	81	81	81	96	103	93	72	80	80
Papua New Guinea	141	142	137	152	138	138	138	187	124	124	124	124
Samoa	32	26	31	14	18	20	21	21	21	21	21	21
Solomon Islands	4	6	8	9	8	11	18	18	18	18	18	18
Vanuatu	7	7	6	2	2	2	2	2	2	2	3	3
Total	266	276	254	258	247	252	275	331	258	237	246	246
Percentage to ASEAN	3.9	3.8	3.2	2.6	2.3	2.2	2.5	2.7	2.0	1.6	1.7	1.7
Production: ASEAN (1,000 m³)												
Indonesia	1,411	1,829	2,415	3,022	3,510	3,501	3,408	4,815	5,269	6,818	6,317	6,317
Malaysia	4,281	4,165	4,002	5,230	5,760	6,019	6,021	6,050	6,312	6,585	7,201	7,283
Philippines	1,061	1,292	1,470	1,609	1,567	1,781	1,626	1,529	1,219	1,200	1,222	1,080
Total	6,753	7,286	7,887	9,861	10,837	11,301	11,055	12,394	12,800	14,603	14,740	14,680
Imports: South Pacific Islands (1,000 m³)												
Fiji	17	14	15	9	9	0	0	0	1	1	1	1
Papua New Guinea	0	0	0	1	0	0	0	0	0	0	0	0
Samoa	1	1	1	1	2	3	2	2	18	16	17	17
Solomon Islands	1	0	0	0	0	0	0	0	0	0	0	0
Vanuata	1	1	1	1	1	1	1	1	0	2	1	1
Total	20	16	17	12	12	4	3	3	19	19	19	19
Imports: ASEAN (1,000 m³)												
Indonesia	0	1	2	1	2	0	0	0	0	0	0	0
Malaysia	76	96	112	158	164	161	220	213	90	84	84	84
Philippines	0	1	0	0	0	0	1	0	0	0	0	0
Total	76	98	114	159	166	161	221	213	90	84	84	84

Tropical Hardwood Supplies 215

APPENDIX 1. Continued.

Sawnwood and Sleepers	1973	1974	1975	1976	1977	1978	1979	1980	1981	1982	1983	1984
Exports: South Pacific Islands (1,000 m³)												
Fiji	3	3	5	8	9	9	7	13	4	6	7	7
Papua New Guinea	30	52	27	51	51	47	47	45	24	21	20	22
Samoa	10	8	3	2	4	4	4	3	2	6	2	3
Solomon Islands	0	0	1	1	2	3	8	8	8	8	8	8
Vanuatu	1	1	1	0	1	0	0	0	1	1	1	1
Total	44	64	37	62	67	63	66	69	39	42	38	41
Percentage to ASEAN	1.5	2.3	1.4	1.5	1.7	1.5	1.1	1.3	0.9	0.8	0.6	0.7
Exports: ASEAN (1,000 m³)												
Indonesia	330	278	395	656	594	757	1,284	1,214	1,182	1,242	1,798	2,200
Malaysia	2,240	2,219	1,914	3,088	2,993	2,827	3,542	3,320	2,808	3,137	3,493	3,501
Philippines	427	284	254	493	455	573	915	742	547	591	728	540
Total	2,997	2,781	2,563	4,237	4,042	4,157	5,741	5,276	4,537	4,970	6,019	6,241
Domestic consumption: South Pacific Islands (1,000 m³)												
Fiji	96	106	82	82	81	72	89	90	90	67	74	74
Papua New Guinea	111	90	110	102	87	91	91	142	100	103	104	102
Samoa	23	19	29	13	16	19	19	20	37	31	36	35
Solomon Islands	5	6	7	8	6	8	10	10	10	10	10	10
Vanuatu	7	7	6	3	2	3	3	3	1	3	3	3
Total	242	228	234	208	192	193	212	265	238	214	227	224
Percentage to ASEAN	6.3	5.0	4.3	3.6	2.8	2.6	3.8	3.6	2.8	2.2	2.6	2.6
Domestic consumption: ASEAN (1,000 m³)												
Indonesia	1,081	1,552	2,022	2,367	2,918	2,744	2,124	3,601	4,087	5,576	4,519	4,117
Malaysia	2,117	2,042	2,200	2,300	2,931	3,353	2,699	2,943	3,594	3,532	3,792	3,866
Philippines	634	1,009	1,216	1,116	1,112	1,208	712	787	672	609	494	540
Total	3,832	4,603	5,438	5,783	6,961	7,305	5,535	7,331	8,353	9,717	8,805	8,523

Source: FAO *Yearbook of Forest Products* 1984.

Chapter 22

Indonesia as a Supplier of Rattan Products in the International Market

SOETARSO PRIASUKMANA

Rattan is classified as a minor forest product. Although it grows on almost all the islands of Indonesia, the species with greatest commercial value are found in Sumatra, Kalimantan, and Sulawesi. About three hundred species grow in Indonesia, of which about fifty have been harvested and marketed (Subiyanto 1986).

In an effort to increase the added value of rattan exports and job opportunities in rattan processing, the government of Indonesia banned export of raw rattan beginning in October 1986 and raised the export tax to 30% on half-finished products until January 1, 1989, at which time exports will be banned. Only finished products—such as furniture, handicrafts, baskets, and mats—will be exported, and no export tax will be charged.

The potential production of sustained-yield rattan has been estimated at about 600,000 tons per year. So far, the maximum production has been 141,993 tons, in 1984 (Table 1). Indonesia has over three hundred large, medium, and small-scale units in the rattan industry, with a total capacity of 337,118 tons per year. In 1985 the total production was 137,738 tons, of which 89,337 tons (65%) were exported to Europe, the United States, Japan, Singapore, Taiwan, Hong Kong, and elsewhere. Domestic consumption in 1985 was estimated at 47,968 tons.

Table 1. Indonesian rattan production and export volume and value, 1981-85.

Year	Production (tons)	Export Volume (tons)	Export Value (U.S. $1,000)
1981	97,515	68,369	73,556
1982	113,431	77,192	83,948
1983	129,154	82,161	86,072
1984	141,993	90,637	93,223
1985	137,738	89,337	97,054

Source: Susatyo (1986).

Production capacity of the three processing levels is shown in Table 2. In the future, the capacity for production of finished products will be expanded as a result of the government policy to promote the export of such products.

Raw and half-finished products exported from Indonesia were W&S (washed and sulfurized) round rattan, core, split, peel (bark), and matting (Table 3). No furniture exports were recorded until 1985.

Table 2. Production capacity of rattan products in Indonesia, 1985.

Type of Industry	Capacity (tons/yr)	%
1. Raw (round) rattan (washed and sulfurized, rough polished, split)	44,942	13.3
2. Half-finished products (core, peel, fine polished)	219,065	65.0
3. Finished products (furniture, handicrafts, mats, webbing, baskets, etc.)	73,111	21.7
Total	337,118	100.0

Source: Khatib (1986).

Table 3. Rattan products exported from Indonesia, 1981-85 (tons).

Product	1981	1982	1983	1984	1985
W&S	49,032	51,274	46,993	52,555	68,490
Core	13,043	17,376	22,581	30,074	10,977
Split	56	26	--	24	--
Peel	5,338	7,298	11,673	7,121	8,460
Matting	490	718	789	862	1,408
Total	67,959	76,692	82,036	90,636	89,335

Source: Subiyanto (1986).

RATTAN CULTIVATION

Rattan belongs to the palm family. Of the several genera that grow naturally in Indonesia, *Calamus, Daemonorops*, and *Korthalsia* are the marketable ones. (See Appendix 1 for individual species names.) Rattan plantations have also been established in Indonesia. In central Kalimantan and East Kalimantan, rattan has been successfully cultivated for more than a hundred years. The major cultivated species are irit (*Calamus trachycoleus*), sega (*C. caesius*), manau (*C. manan),* tohoti *(C. irops),* semambu (*C. scipionum*), and pulut (*Calamus* spp.).

The government of Indonesia, realizing the export potential of rattan, has begun to pay attention to its cultivation, including promotion of research on seedlings, inventory of rattan species that can be cultivated, silvicultural systems, pest and disease control, and so forth.

Trial plantations were started in 1981 in Java, Sulawesi, and Sumatra by the Department of Forestry. The state company Perhutani plans to grow rattan in production and conservation forests of up to 100,000 hectares in Java, and so far has planted about 1,050 hectares. The state company P. T. Inhutani I plans to grow rattan in East Kalimantan, and P. T. Inhutani III in central Kalimantan. The species to be planted are *Calamus irops, C. robustus, C. simphysipus, C. manan, Daemonorops hysterix*, and others.

Research and development on rattan planting and processing are also under way. A three-year grant was established in 1985 by IDRC (International Development Research Centre) and may be extended several more years.

RATTAN PRODUCTION AND INTERNATIONAL TRADE

As was mentioned before, potential annual production in Indonesia of rattan—both natural and cultivated—has been estimated at about 600,000 tons per year. A breakdown of this estimate by province is shown in Table 4.

Table 4. Potential annual production of rattan in Indonesia.

Province	Production (tons/yr)		Total
	Natural*	Cultivated**	
Aceh	45,000	--	45,000
Riau	2,840	30	2,870
North Sumatra	6,000	--	6,000
West Sumatra	34,000	90	34,090
Jambi	6,900	--	6,900
Bengkulu	23,100	--	23,100
South Sumatra	5,000	--	5,000
Lampung	24,000	--	24,000
West Java	--	960	960
West Kalimantan	92,500	--	92,500
Central Kalimantan	24,000	300	24,300
South Kalimantan	7,000	24,600	31,600
East Kalimantan	11,650	7,000	18,650
North Sulawesi	87,000	60	87,060
Central Sulawesi	18,400	--	18,400
South Sulawesi	150,000	360	150,360
West Nusa Tenggara	36,000	--	36,000
Total	573,390	33,400	606,790

*Subiyanto (1986).
**Estimated by the author.

The flow of production and trade in rattan products from raw material to consumer is shown in Figure 1.

Based on statistical data of ITC Geneva, the value of imported rattan products from the main exporting countries has increased as shown in Table 5. The value of these products increased from U.S. $12,654.21 million in 1981 to $15,156.71 million in 1985, or 22.8%. The value of other imported finished products increased only 2.8% from 1981 to 1985.

Indonesia is the dominant exporting country of rattan, supplying about 75% of the world market (see Table 6).

With the new government policies on rattan trade, a new era of the rattan furniture industry in Indonesia has begun. According to Bunke (1986), based on an export value of rattan in 1985 of U.S.$97 million (Table 1), if this raw material were converted into finished furniture, the value would amount to U.S. $350 to $400 million. This would be equal to a volume of 65,000 to 70,000 (20 foot) containers per year of furniture, requiring 130 to 150 factories with 5,000 to 6,000 m^2 production area and about 500 workers each. Such production would greatly stimulate other industries, transportation, port activities, and so forth.

Many countries produce rattan furniture, but the raw material is mostly from Indonesia. Since this type of furniture is sold in all developed countries and many developing ones, there is a steady

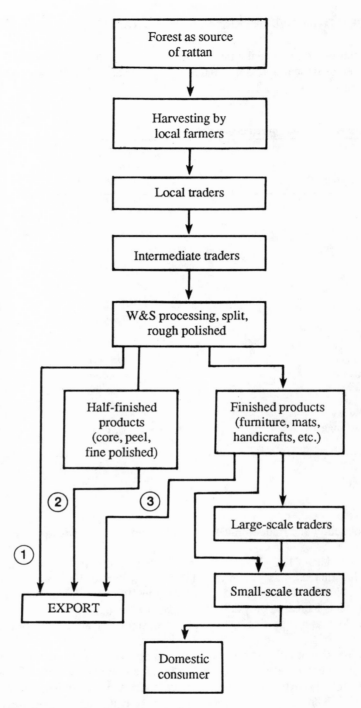

Figure 1. Flow of production and trade of rattan products.
W & S = washed and sulfurized rattan.

Table 5. Imported rattan products from the main exporting countries (primarily Indonesia, China, Singapore, Hong Kong, and Taiwan).

Commodity	Value ($ million)	
	1981	1985
Plaiting materials	124.80	113.45
Plaits, plaited products	135.33	115.43
Baskets, etc.	911.33	927.33
Subtotal	1,171.46	1,156.21
Chairs, seats, and parts	7,148.28	9,976.80
Furniture and parts	4,334.47	4,023.76
Subtotal	11,482.75	14,000.56
Total	12,654.21	15,156.77

Source: Algamar (1986).

Table 6. Indonesia's share of world trade of raw rattan, 1980 (1,000 tons).

Exporting Country	Importing Country					
	Developed Countries	Hong Kong	Singapore	Taiwan	Total	Share (%)
Indonesia	11.35	41.64	19.16	9.11	81.26	75
Philippines	7.81	0.11	0.06	--	7.98	7
Thailand	6.32	2.46	0.16	--	8.94	8
Malaysia	5.99	0.26	0.26	1.15	7.66	7
Mexico	0.57	--	--	--	0.57	0
Others	1.91	--	--	--	1.91	2
Total	33.95	44.47	19.64	10.26	108.32	100

Source: Subiyanto (1986).

demand for it every year. Considering that the worldwide demand for rattan furniture reached a value of U.S. $14,000 million in 1985 (Table 5), and Indonesia has the major supply of the raw material needed for this industry, the opportunity for Indonesia to enter the international rattan furniture market on a large scale is excellent.

REFERENCES

Algamar, K. 1986. The position of Indonesian rattan for the international market. Rattan National Workshop, Department of Forestry, Jakarta, December 15-16, 1986.

Bunke, K. 1986. The development of the rattan furniture industry in Indonesia. Rattan National Workshop.

Khatib, Z. 1986. Management and development of the rattan processing industry. Rattan National Workshop.

Priasukmana, S. 1986. The economic valuation of commercial rattan cultivation in East Kalimantan. Rattan National Workshop.

Subiyanto, 1986. Investment opportunity for rattan business. Rattan National Workshop.

Susatyo, A. 1986. The experience of rattan cultivation in practice. Rattan National Workshop.

Suwanda, A. 1986. Rattan from Indonesia as an export commodity. Rattan National Workshop.

APPENDIX 1. Marketable rattan species from natural growth.

Botanical Name	Local Name
Calamus axilaris	Rotan air
Calamus boniluris	Ronti
Calamus caesius	Sega (taman)
Calamus cejocaulic	Jermasin
Calamus dymphysipus	Ombal
Calamus hisdulus	Bulu
Calamus irops	Tohiti
Calamus javensis	Chili, cabang
Calamus matanansis	Sabut
Calamus manan	Manau
Calamus ornatus	Batang
Calamus regendaris	Ranting, cabang
Calamus ranci jugus	Andaru
Calamus ruvidus	Lilin
Calamus scipionum	Semambu
Calamus trachycoleus	Irit
Daemonorops diaco	Jernang
Daemonorops melanochaetes	Seel, manis
Daemonorops roburtus	Batang
Daemonorops tricharous	Getas
Daemonorops verticillaris	Cincin
Korthalsia rigida	Paku, manau merah, dahan
Korthalsia scaphigera	Udang tali
Korthalsia zepplii blune	Inawai

The Example of One State
in a Global Economy

Chapter 23

Importance of the Forest Products Industry to the State of Washington

JOHN C. ANDERSON

No state in this country has a greater interest in the forest products industry—or in international trade—than Washington State. This symposium is therefore right on target and also most timely. In the following comments I would like to discuss the importance of the industry to this state, the state's potential for trade in forest products, and what we in Washington are doing to enhance the growth of the industry and its international markets.

Washington—the Evergreen State—is almost synonymous with forests. We in the state recognize the great importance of the forest resource to our collective livelihood, and the beauty and recreational benefits forests lend to the enrichment of our lives. Over time the forests have been good to us economically, providing a ready industry to the early settlers and then supporting many generations of Washingtonians in many areas of the state.

The health of the forest products industry is important to Washington State government, since the industry is a large part of the state's economy, and the state's economy is the engine that generates nearly all public revenues. That is the rationale behind state government's involvement to promote the industry's domestic and international exports, and to encourage investment in forest products enterprises.

In 1985 the forest products industry generated $7 billion in gross business income and exported $1.4 billion in goods to international markets—9% of total exports. With 53,000 jobs, the wood products industry is number two in the employment category—right behind aerospace—producing an annual payroll of about $1.5 billion. The high water mark for employment came in 1977, when more than 71,000 people were working in the wood and paper industries. The low point appears to have been reached in January 1986—with an employment of 52,000—caused by the housing and interest-rate recession.

Recent newspaper articles have reported resurgent earnings by such companies as Weyerhaeuser, Boise Cascade, and Willamette Industries. This offers encouraging evidence that the Northwest forest products industry is emerging from the period of painful restructuring with renewed vigor. In the past few years the industry has worked to control labor, distribution, and transportation costs, thus improving its productivity. Today the industry is profitable, and competitive against world standards. Lumber output during 1986 reached 87% of 1978 peak year production with one-third fewer workers. There should be no doubt that forest products will continue to play a major role in the Washington State economy far into the future.

Washington State is more dependent on international trade and investment than any other state and than most foreign countries. Washington's two-way international trade flow in 1985 totaled nearly $34 billion. Of that amount, 71% ($24 billion) was in trade with Pacific Basin countries. Trade with Japan alone—Washington's largest trading partner—amounted to $13 billion, or nearly 40% of the state's total world trade.

Washington is the principal U.S. gateway for Pacific Basin trade between Asian economies and the 100 million Americans living in the U.S. northern tier states. We are one of the most internationalized states in the country.

The housing and interest-rate recession of the early 1980s taught the wood products industry a painful lesson: despite its exports, the industry had overrelied on domestic markets. All our traditional industries must come to recognize that increasingly we are living, working, and competing in a global economy. In the future the opportunity for this industry lies in broader international trade.

Although the United States has increased wood exports in this decade, today it accounts for only 20% of global wood exports. Even though Washington statistically accounts for a large share of U.S. forest products exports, there is little question that the state's wood industry has enormous potential for significant export-based expansion. In fact, Washington has unique advantages which should enable it to exploit its export potential with Pacific Basin markets in particular: (1) abundant forest resources, (2) an available and experienced work force, (3) a highly productive industry, (4) available plant capacity for expanded production, (5) a highly developed transportation system to move exports to market, and (6) a strategic location from which to export to the rapidly expanding markets of the Pacific Basin.

Japan is clearly our strongest market, accounting for 56% of Washington's forest products exports. Some 60% of exports to Japan are in the form of logs. Finished wood products are expected to capture an increasing share as producers adjust to Japanese product specifications. There is enormous potential for increasing exports to Japan, considering that 1985 exports—valued at $779 million—were only half what they were in the peak year of 1979. Increased Japanese government spending on public housing in 1986 helped increase housing starts to 1.3 million, and industry groups are predicting 1.5 million housing starts for 1987. In addition, the Japanese are modifying building codes in ways that would promote remodeling and expansion of existing wooden houses, allowing for greater use of U.S. wood. The demonstration Summit House, which went on display in Tokyo in May 1986, so far has been seen by 23,000 people—most of them in the building professions. The Japanese Ministry of Construction is even proposing the creation of a "Wooden House Development Office" to promote the use of wooden houses in Japan.

The People's Republic of China, receiving 17% of Washington's exports of wood products, is the state's second largest export market, in 1985 accounting for $244 million. Because of its current pace of modernization, and demand for wood far beyond the resources of its domestic forests, China should continue to be a good export customer.

South Korea is the state's third largest export market, accounting for more than $141 million in wood products exports in 1985 (about 10%). Canada is the fourth largest purchaser, importing $118 million worth in 1985 (about 8%).

Taken together, these four Pacific Basin countries account for at least 90% of the state's wood exports.

EXPANSION OF TRADE

Several organizations recognize the state's great potential to widen forest products trade, and they are working in partnership toward that end. I will briefly describe some of these organizations, because they illustrate the cooperative nature of our approach to trade and economic development in Washington State.

Team Washington is the state's umbrella economic development organization, made up of state government agencies and private sector organizations and individuals working together across the state. Each of the thirty-nine counties has designated one public or private organization to participate in Team Washington, and receives annual funding to maintain a full-time economic development office. All of these organizations are capable of working with the Department of Trade and Economic Development on a day-to-day, continuing basis on international trade and investment matters. In addition, several statewide nonprofit organizations with international trade development interests also participate in Team Washington.

The Department of Trade and Economic Development has a leadership role in Team Washington. The Department provides both financial and staff support to assist the forest products industry in many ways, especially by encouraging wood products exports and attracting investment in the forest-based industry of the state.

Trade

The Department has been a major supporter of CINTRAFOR (Center for International Trade in Forest Products), the sponsor of this symposium. CINTRAFOR's key role is to perform export market analysis and research. The Department supported CINTRAFOR both before and since its establishment in 1984 in the College of Forest Resources at the University of Washington. Tom Waggener, CINTRAFOR's director, is a distinguished professor of forest economics and policy at the university and an enthusiastic participant in Team Washington.

CINTRAFOR receives state funds through the Department of Trade and Economic Development as well as from private industrial firms and the U.S. Forest Service. Among CINTRAFOR's early accomplishments are a graduate degree program emphasizing international trade in forest products, a statistical base of market and product information, which is now operational, and new research on specific problems and opportunities in international trade.

CINTRAFOR's executive board includes executives of major private sector forest products companies as well as state agency heads, University of Washington officials, and U.S. Forest Service professionals.

The Evergreen Partnership, a new nonprofit corporation, was formed by state agencies in partnership with the private sector in June 1986. Its goal is to actively help the industry develop and expand international and domestic markets for Northwest forest products. The Department of Trade and Economic Development actively supports the Partnership with funding and other support.

The organization is directed by Tom Westbrook, previously president of the Pacific International Trade Corporation of Olympia. It has signed up active and associate corporate and public sector members, who provide the financial support essential to conduct an active development and marketing program to increase the sales of Northwest forest products in domestic and world markets.

The Evergreen Partnership provides marketing in a broad sense, specifically to increase awareness throughout the world of Washington State's tremendous breadth of wood products capability; an informational network, offering information on suppliers of forest products; and trade leads and product requests. The Partnership acts as a clearinghouse for information flowing in both directions.

Here again is a cooperative effort among state government agencies—including the Department of Trade and Economic Development, the Department of Agriculture, the Department of Natural Resources, and the University of Washington—and private sector forest products companies, including but not limited to the industry's major players.

All of the agencies just mentioned cooperated in the production of a *Forest Products Trade Directory*—an ambitious attempt to list all major organizations and companies that make up Washington's wood products industry and supporting agencies. This new tool is an important step in building awareness in domestic and foreign markets of the diverse components of the state's forest products industry.

Other cooperative efforts are taking place. In November 1986 a Pacific Basin Housing Conference and Exhibition, hosted by the congressional delegations of Washington and Oregon and by both states' governors, was held in Portland, Oregon and Vancouver, Washington to focus on the serious housing needs of countries that are or soon will be prime potential markets for Northwest forest products. Those countries are China, India, Japan, Micronesia, Papua New Guinea, the Philippines, the Republic of Korea, and Sri Lanka. The conference was attended by top ranking housing ministers from the eight countries, and provided an opportunity to develop strategies to promote expansion of wood exports and housing technology from our region to important developing markets.

Recently in Seattle, state agencies, along with CINTRAFOR and the Evergreen Partnership, sponsored a conference on the factory-crafted housing industry of Sweden. The two-day conference introduced to Washington's own factory-built housing firms the technology of Sweden's highly successful industry, and offered the opportunity to serve domestic U.S. and foreign markets.

Identifying new and expanded markets for Northwest forest products was a principal goal of Team Washington economic missions to Asia and Europe in the fall of 1986. This mission traveled to Japan, Taiwan, and the Republic of Korea. Among the seventy-five members of the three-week mission, led by Governor Booth Gardner, were representatives of the state's forest products industry, including Tom Westbrook of the Evergreen Partnership and senior executives of the Quadrant Corporation and the Weyerhaeuser Company. Mission members also included employees of Washington State ports, state trade and investment professionals, and Team Washington organizations across the state.

The previous Team Washington mission to Asia, in 1985, included Dean David Thorud, who traveled with Governor Gardner and other members throughout China to explore new market opportunities for the forest products industry.

In 1986 Team Washington also launched its first economic mission to Europe, visiting eight countries. The forest products contingent included a director of the Evergreen Partnership, Tom Waggener of CINTRAFOR, and forest products industry trade and investment specialists from the Department of Trade and Economic Development. This experienced group worked to increase awareness of Washington's wood products industry and explored opportunities for trade, investment, and joint ventures involving forest products.

Investment

Another economic development program—Investment—is actively encouraging and recruiting investment in new plant facilities to produce new and diversified products *not* currently produced in Washington, but utilizing the existing forest base. Most of this investment so far is in products for export. I'll mention two examples.

In July 1986 the Team provided significant site location consulting assistance to the United States Waferboard Corporation, which decided to locate a $45 million waferboard plant near McCleary in Grays Harbor County. A significant portion of production will eventually be exported to U.S. domestic and overseas markets. This project demonstrates the state's ability to meet essential investment specifications for forest products investments, including a trained labor force, available wood resources, and proximity to a modern transportation system.

For the past two years Team Washington has been working with the Ponderay Paper Company, which is planning a major paper manufacturing plant in Pend Oreille County. When it is built, the plant is expected to supply several out-of-state newspaper operations.

The Investment Division of our Department of Trade and Economic Development is currently seeking domestic and foreign investment in expanding forest products firms, and in acquisitions of inactive forest products facilities. Sectors that have promising potential include: reconstituted wood products, which use residual and underutilized wood resources; pelletized fuel, which uses existing technology to make a clean burning product that meets wood pollution standards; remanufactured lumber products, including factory-fabricated door and window frames and moldings; pulp and paper products, currently enjoying strong demand; and plywood and lumber products, including manufactured housing, which should become more economically viable with automated mills and lower labor costs.

In all of its trade and investment promotion efforts, Team Washington has the goal of expanding the state's forest products exports to which value has been added, through processing or remanufacturing—to maximize job creation in the state.

CONCLUSION

The forest products industry is crucially important to Washington State's economy, and increased trade offers an important opportunity to broaden the markets for the industry's products—especially processed or remanufactured products. We in Washington, on both public and private fronts—and working in partnership—recognize that opportunity and are moving to realize it through focused market research and promotion conducted by energetic new organizations, through cooperative economic missions, and through investment recruiting. We believe we are on the right track and we expect to be successful.

Contributors

Erik T. Anderson
College of Forest Resources   AR-10
University of Washington
Seattle, Washington   98195

John C. Anderson
Director, Washington State Department of
    Trade and Economic Development
101 General Administration Building
Olympia, Washington   98504

Philip A. Araman
USDA Forest Service
P.O. Box 152
Princeton, West Virginia   24740

James S. Bethel
College of Forest Resources   AR-10
University of Washington
Seattle, Washington   98195

Keith A. Blatner
Department of Forestry and Range
    Management
Washington State University
Pullman, Washington   99164-6410

David G. Briggs
College of Forest Resources   AR-10
University of Washington
Seattle, Washington   98195

Arnaldo Jelvez Caamaño
Universidad del Bíobio
Casilla 5-C, Concepción
Chile

Sun Joseph Chang
Department of Forestry
University of Kentucky
Lexington, Kentucky   40546-0073

Theo Erfurth
Orthstrasse 12
8000 Muenchen 60
Federal Republic of Germany

Fred A. Flynn
Fiji Forest Industries Limited
12 Gillard Place
North Ringwood
Victoria 3134, Australia

Jia Ju Gao
Jiangsu Wood Industry Society
137 Jian Kang Road
Nanjing, PRC

Alberto Goetzl
National Forest Products Association
1250 Connecticut Ave. N.W., Suite 200
Washington, D.C.   20036

Robert L. Govett
Department of Forest Products
University of Idaho
Moscow, Idaho   83843

Jay A. Johnson
College of Forest Resources   AR-10
University of Washington
Seattle, Washington   98195

Jan G. Laarman
Department of Forestry, Box 8002
North Carolina State University
Raleigh, North Carolina   27695-8002

Robert H. Leicester
CSIRO Division of Building Research
P.O. Box 56, Highett, Victoria 3190
Australia

Stephen M. Lovett
National Forest Products Association
1250 Connecticut Ave., N.W., Suite 200
Washington, D.C.   20036

Borg Madsen
Department of Civil Engineering
University of British Columbia
Vancouver, British Columbia
Canada   V6T 1W5

T. M. Maloney
Wood Materials and Engineering
    Laboratory
Washington State University
Pullman, Washington 99164-1806

H. M. Montrey
Forest Products Laboratory
USDA Forest Service
One Gifford Pinchot Drive
Madison, Wisconsin 53705-2398

Marcio A. R. Nahuz
Instituto de Pesquisas Tecnológicas do
    Estado de São Paulo
Caixa Postal 7141
São Paulo, Brazil

Yuji Noga
Issiki Architects & Partners Company Ltd.
3–9–5, Yoyogi, Shibuya-ku
Tokyo, Japan 151

T. J. Peck
FAO/ECE Agriculture and Timber Division
Palais de Nations
Geneva 10 CH-1211
Switzerland

Soetarso Priasukmana
Balai Penelitian Kehutanan Samarinda
Jl. Ir. Juanda No. 206
Samarinda, Indonesia

David L. Rogoway
American Plywood Association
P.O. Box 11700
Tacoma, Washington 98411

Gerard F. Schreuder
College of Forest Resources   AR-10
University of Washington
Seattle, Washington 98195

W. Ramsay Smith
College of Forest Resources   AR-10
University of Washington
Seattle, Washington 98195

C. K. A. Stieda
Forintek Canada Corporation
6620 N.W. Marine Drive
Vancouver, British Columbia
Canada   V6T 1X2

Marshall R. Turner
Western Wood Structures, Inc.
20675 S.W. 105th Street
P.O. Box 130
Tualatin, Oregon 97062

Yeo Youn
College of Forest Resources AR-10
University of Washington
Seattle, Washington 98195

Index